T0133396

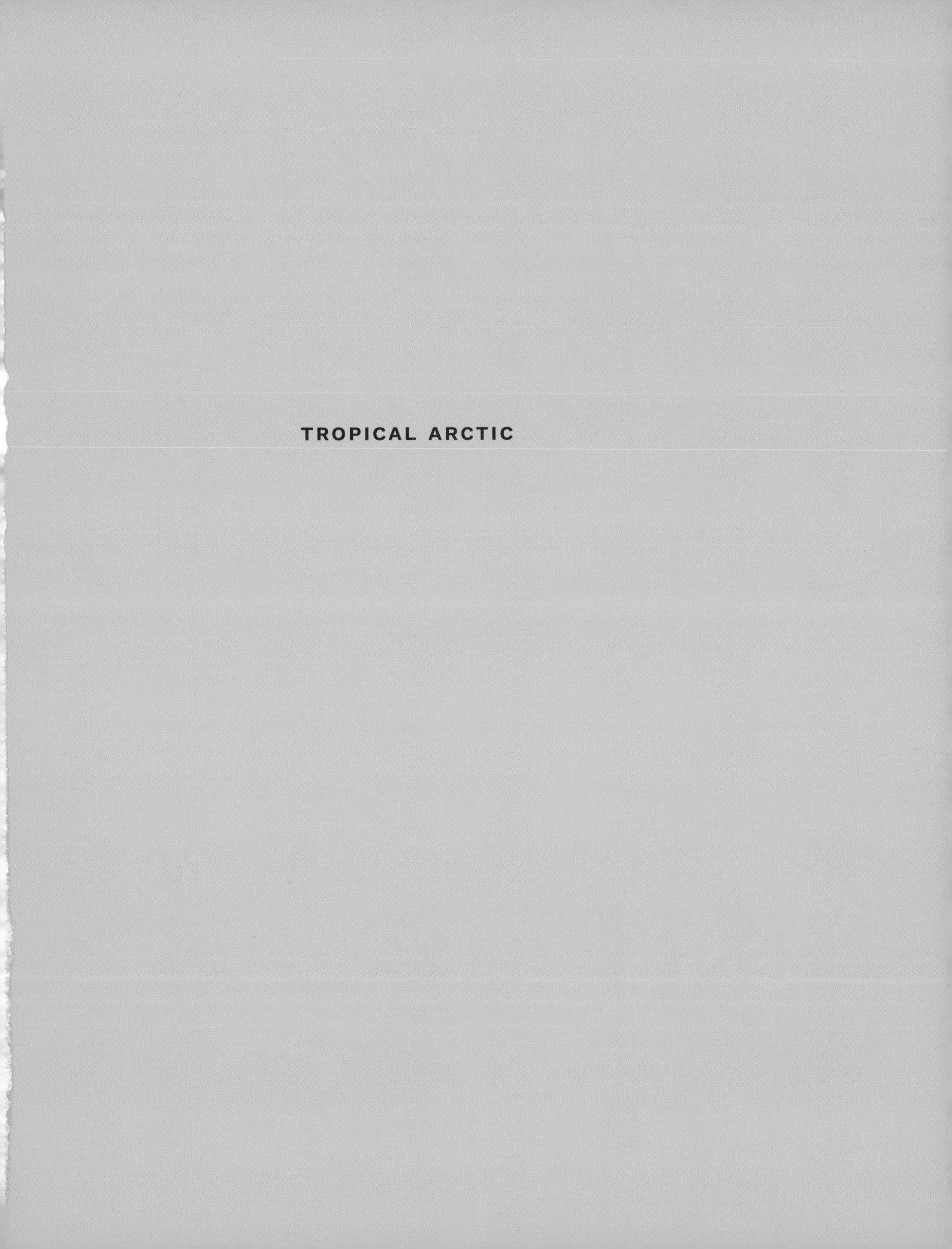

TROPICAL ARCTIC

JENNIFER C. McELWAIN,
MARLENE HILL DONNELLY,
AND IAN J. GLASSPOOL

Tropical Arctic

Lost Plants, Future Climates, and the Discovery of Ancient Greenland

THE UNIVERSITY OF CHICAGO PRESS CHICAGO & LONDON

The University of Chicago Press, Chicago 60637
The University of Chicago Press, Ltd., London

Published 2021
Printed in China

30 29 28 27 26 25 24 23 22 21   1 2 3 4 5

ISBN-13: 978-0-226-53443-5 (cloth)
ISBN-13: 978-0-226-53457-2 (e-book)
DOI: https://doi.org/10.7208/chicago/9780226534572.001.0001

Library of Congress Cataloging-in-Publication Data

Names: McElwain, Jennifer C., author. | Donnelly, Marlene Hill, illustrator. | Glasspool, Ian J., author.
Title: Tropical Arctic : the science and art of lost landscapes / Jennifer C. McElwain, Marlene Hill Donnelly, and Ian J. Glasspool.
Description: Chicago ; London : The University of Chicago Press, 2021. | Includes bibliographical references and index.
Identifiers: LCCN 2020058553 | ISBN 9780226534435 (cloth) | ISBN 9780226534572 (ebook)
Subjects: LCSH: Paleobotany—Greenland. | Paleobotany—Triassic. | Paleobotany—Jurassic. | Paleoecology—Greenland. | Paleoecology—Triassic. | Paleoecology—Jurassic.
Classification: LCC QE922 .M37 2021 | DDC 561/.19982—dc23
LC record available at https://lccn.loc.gov/2020058553

♾ This paper meets the requirements of ANSI/NISO Z39.48-1992 (Permanence of Paper).

Paleobotany has illuminated our knowledge of Earth as it was, and may yet become, in ways that few other disciplines can.

This book is dedicated to all, whatever their role (artist, scientist, volunteer, collections manager), who have contributed to this journey of understanding.

Contents

Illustrations

Preface

No human has ever seen the landscapes that you are about to explore. Reconstructed from research on plant fossils, the illustrations in this book present a distant but real Earth through the perspectives of two scientists and an artist. Together, we will witness a 10 million-year period of our planet's past, when greenhouse gases rose precipitously, global temperatures soared, and the vegetation of the world responded. You will see how the sciences of paleobiology, geology, and paleobotany, together with scientific illustration, can bring fossils to life, so that we can better understand the consequences of climate change.

The three of us didn't set out with such goals. This book began in 2002 with a simple but probing question from Marlene, the artist: "What color were the leaves before they became fossilized?" Botanist Jenny's gut reaction was that this was an inappropriate question and, from a scientific perspective, an impossible one to answer. Upon reflection, she soon realized the value of the question—both for the answer and for collaboration between our disciplines. Determining the key characteristics of modern leaves that influence their color could also give us broad clues to their biological function. At the most basic level, the perceived color of any object is a function of the light that object reflects and the light it absorbs. When the sun shines "white" light—a mixture of many frequencies of electromagnetic radiation—on a "green" leaf, we see green simply because the leaf reflects the frequencies that our eyes identify as green and absorbs the others. Why green? The reason is in part that to avoid the greater energies associated with shorter wavelengths that risk genetic damage, many leaves evolved to absorb visible red light and convert that light into energy. But leaves do this in many ways—and there are many shades of green. In order to assess and incorporate a diversity of greenness into the lost landscapes that we will reveal over the following pages, we used the microscopic clues preserved in and on fossil leaf surfaces. We investigated threadlike fossilized hairs and waxy leaf layers to infer the leaf's levels of reflectance and to estimate the color that it displayed when it was living, over 200 million years ago. Step by step over a decade, we identified the characteristics of leaf structure to answer Marlene's probing question, to determine not only the color of particular leaves, but of the lost landscapes of East Greenland at the end of the Triassic and the dawning of the Jurassic. When we completed our work, we became the first people to "see" East Greenland as it was millions of years ago, during what scientists call a "hothouse Earth" period, when the planet was ice-free and uninhabitable by us.

As an artist who loves the land, Marlene found the idea of re-creating an ancient landscape both daunting and exciting. Science requires accuracy, but to accomplish our goal, the final work also needed light, life, and depth. Staring at a fossil and a stark photo of its Arctic Greenland site produced no inspired vision of the original living forest. It was like trying to reconstruct a conversation from a single whispered word. Our rich trove of fossils eventually told us more, speaking of a world quite the opposite to today's Greenland. Lush tropical plant fossils, many and varied, shouted heat and humidity. The rocks hosting them mumbled of water and mud or sand. Over the following years, we gave the fossils the greatest attention and respect in order to hear everything they had to say and to interpret what they were saying as accurately as possible. We heard them ever more profoundly, in unfolding layers of revelation, listening to a story of deep time. Reading this story, you will hear 24 recovered conversations—or more accurately, see 24 views of East Greenland's lost landscapes, each carrying its own message. Over chapters 2–4, they combine to tell, in the final three complete paintings, a story about environmental upheaval, a mass extinction event, the resilience of plant life, and the transition between the Triassic and Jurassic periods. We invite our readers to use the appendix at the back of this book to delve deeper into the paleobotany of Greenland's lost plants as the story of our discovery of ancient Greenland unfolds.

Presenting such a comprehensive picture was unusual for Ian, as it would be for any paleobotanist. We most frequently find fossil plants as small, fragmentary, isolated parts—imagine the forest floor in autumn, or following a storm, when it's strewn with leaves, twigs, catkins, cones, and fruits. For a paleobotanist, it's often daunting to try to visualize these isolated fragments as whole plants and to understand their position and role in a broader ecosystem. How do you piece together which isolated leaves were attached to which isolated twigs and thence to which isolated cones? When you lack a picture of the whole plant, you don't even begin to imagine that plant on a broader landscape. However, that's what these reconstructions demanded. Only through collaboration has it been possible for us to attempt to answer some of the searching questions these Greenland fossils have posed as we sought to reconstruct three ancient landscapes from either side of one of Earth's great mass extinction events. Eighty-five percent of plant species in Greenland went extinct across the Triassic-Jurassic boundary, and globally, about a quarter of all animal families suffered extinction, making the Triassic-Jurassic the third greatest extinction event in Earth history. We have followed in the footsteps of geologist Charles Lyell, who, during the Scottish Enlightenment, observed that "the present is the key to the past." Since at least the time of botanist Albert Charles Seward in the 1930s, paleobotany has been influenced by a derivative of this principle, that "the past is the key to the present." This principle suggests that in order to study the potential impacts of modern climate change on Earth's ecosystems, we must understand the fossil record and global biota. We hope this book will demonstrate how landscapes of the past can be used to inform the present and to predict the health of future ecosystems in a world in the midst of a climate and biodiversity crisis.

1 A Journey into the Past

A small herd of musk oxen gallops below as we fly in a helicopter over the vast, stunted tundra landscape. The tundra is a unique ice-adapted vegetation and is considered one of the most threatened by current climate change. It will be thawed out of existence by rising temperatures if global climate warming remains unchecked. Dry hummocks—groups of tiny mounds—and sodden wet hollows—small holes—stretch below as far as our eyes can see. The plants that make up this treeless vegetation are low in species diversity, but vary in character and composition between the wetter and drier areas, resulting in a great floristic mosaic. Shades of brown, orange, and dull green sweep to the horizon. Billowing white flashes mark patches of bog cotton and remnants of snow unlikely to melt even in the high summer of the Arctic. All tundra vegetation today is low, growing in soils that are permanently frozen except for the topmost active layer. There are no trees in the modern Arctic landscape—it is too cold—but woody plants abound, including dwarf willow and birch, growing flattened to avoid wind chill. Another feature of the east coast of Greenland that we are skirting is the high proportion of bare land, where soils have yet to become established and vegetation cover is poor or absent. In these most recently deglaciated environments, only mosses and lichens survive. In winter, snow cover obscures these patches, and the active soil layers refreeze; in summer, however, they lie exposed, demonstrating the harshness of these high polar landscapes despite seasonal thaw.

In July 2002, more than 70 years after British paleobotanist Tom Harris's

FIGURE 1.1. Photos of modern Jameson Land landscape, flora, and fauna. *Clockwise from top left*: landscape around Nerlerit Inaat Airport; cottongrass; moss campion; wintergreen; musk ox skull; Arctic willow; willowherb; moss sporophytes.

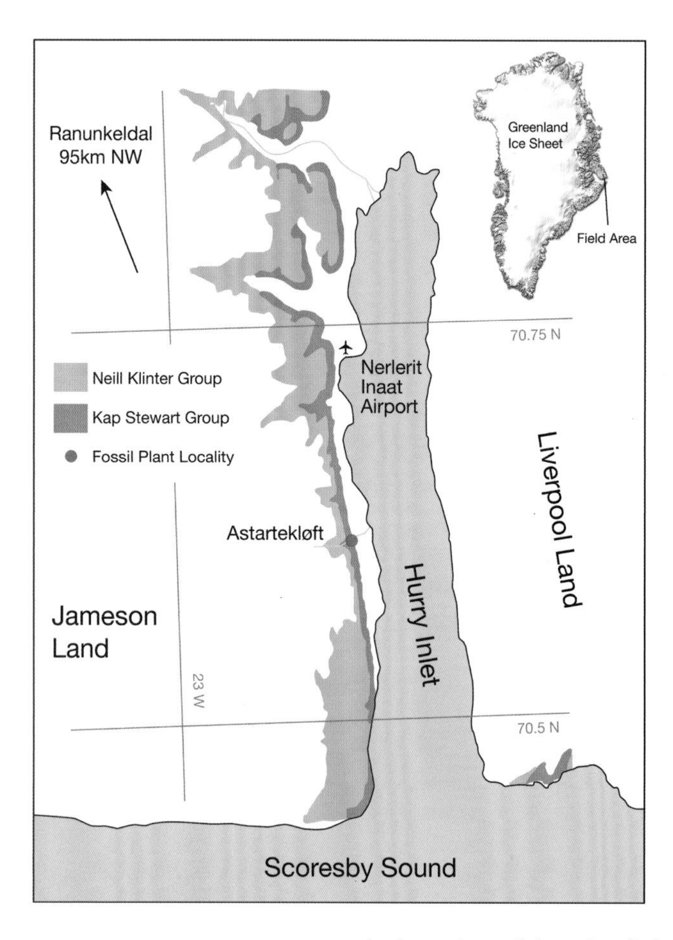

FIGURE 1.2. Map of Jameson Land, showing expedition localities and geology.

pathfinding expeditions to the region, we were part of a National Geographic–funded expedition to discover and collect new plant fossils lying untouched, for hundreds of millions of years, in ancient sediments below East Greenland's frozen tundra. We were a small group of researchers from Chicago, Copenhagen, Oxford, and Bucharest, almost complete strangers to one another, who had rendezvoused in Reykjavík with the ambitious plan of reconstructing Greenland's lost floras as they transitioned through the third greatest mass extinction event in history at the close of the Triassic period, about 200 million years ago.

A prop-plane flight from Reykjavík brought us into our Jameson Land expedition base, Nerlerit Inaat. The Nerlerit Inaat airport is one of the most beautiful and remote air bases in the world. There are no roads linking it to human populations in the surrounding Sermersooq area (whose name translates from the Inuit as "place of much ice"). The only road, a runway, leads into Earth's deep past, as it is flanked to the west by high cliffs that span the boundary of the Triassic (252–201 million years ago) and Jurassic (201–145 million years ago) periods.

From Nerlerit Inaat, the helicopter took off with our party of five, along with four weeks of food, camping, and collecting supplies. Our expedition had truly started. We were headed to our first field site, Ranunkeldal, on the Jameson Land peninsula, one of our northernmost sites. Finn Surlyk, from the Geological Survey of Denmark and Greenland, navigated for the pilot by counting the streams cutting through the Late Triassic– and Jurassic-aged cliffs as we traveled north. We followed the invisible contours of the once formidable Kap Stewart Lake that occupied the Jameson Land peninsula during this interval and beyond. To the untrained eye, the ancient lake margins are invisible, but they have been extensively mapped through repeated studies of the rock sediments by Finn and his colleagues. This fundamental knowledge allowed us to interpret the paleo-landscape, now a ghostly relic in the great cliffs and remarkable outcrops of the Jameson Land region.

Our aim as a group of geologists and paleobotanists on this first expedition was to build a detailed scientific visualization of what the East Greenland landscape looked like in the Triassic and Jurassic. We were particularly interested in studying 208–201 million-year-old Rhaetian Age and 201–199 million-year-old Hettangian Age sediments from the very latest Triassic and the very earliest Jurassic, respectively. These rocks and sediments contained the biological record of one of Earth's "big five" mass extinction events, a biological crisis thought to have been triggered by a global warming of 6°C (10.8°F) or more. We wanted to understand what species and vegetation occupied the shores of Kap Stewart Lake before, during, and after the Triassic-Jurassic mass extinction event. What was the prevailing climate at this time and could we draw any parallels between the events of the past and our current biodiversity and climate emergencies? Our plan was to extract this information from fossil plants preserved in the sediments and rocks making up the jutting cliffs that now stood as our helicopter's navigational guide northward to our first field site.

We chose a flat plateau site to land the helicopter and set up camp, due south of Ranunkeldal, which translates from the Danish to "Buttercup Valley." Finn recommended the elevated site because valley camping in the Arctic is too cold. Sea mist rolls in just after midday and the rock cliffs prevent the daytime sun from penetrating the valley depths. Camp at 445 meters elevation consisted of a bedroom tent for each expedition member and one old-school canvas kitchen tent that we used for indoor cooking, dining, wrapping, and identifying fossil plants as well as writing up the day's field notes. Each member of our group had defined roles—both scientific and more mundane day-to-day camp duties. We soon found that we represented a mix of experience, from completely uninitiated camper to seasoned fieldworker.

RANUNKELDAL CAMP

Nothing is quite like waking up in the red early morning glow of your first expedition morning in Greenland. The warmth of the sun heats your tent, and a gush of bracing pristine, crisp air rushes in as you unzip the door. The landscape appears utterly desolate when viewed at a macro scale, but soon begins to reveal its dwarf splendor and diversity as you investigate in more detail. Rock surfaces are encrusted with orange lichens; mosses abound underfoot where the melting glaciers supply a constant flow of water and nutrients. Behind the Ranunkeldal camp was what we referred to as our "plant glacier"—a tongue-like mass of Arctic alpine plants taking on the shape of an advancing mountain glacier. It marked the shape and path of a former glacier that had since retreated due to our currently warming climate, leaving behind new ground for plants to colonize. This type of landscape-scale alteration is happening across Greenland today and will pick up pace in the future due to global climate change. Due to a phenomenon known as polar amplification, the high latitudes will warm at a faster pace, and by a greater magnitude, than other regions on Earth. This phenomenon brings us back to the discussion of absorption and reflection of light with which this book opens. In general, darker materials absorb more light and get warm more quickly; light and white materials reflect more light and stay cooler. This is why you will see white roofs on houses (and buses) in warm climates. It is also part of the reason why the poles of our planet, covered in ice and clouds, remain cool. This creates what scientists call a feedback loop; as the planet warms, that ice melts, reflecting less sunlight; more clouds form, trapping that heat; and more heat moves via oceanic and atmospheric circulation to the poles—causing the planet to warm further. When we take this knowledge into the deep past, it means that our quest to reconstruct the past landscapes of Arctic Greenland is particularly relevant. High latitudes are sensitive indicators of the state of the entire global climate, both in the past and today.

Our camp quickly took on a typical daily morning rhythm before we headed out on fieldwork each day. We brewed strong coffee on the stove each morning, made hearty soup for lunch, and reconstituted milk for cereal from stream water and dried powder. Jenny especially enjoyed wedges of Parmesan cheese and cured sausage with crackerbread. We also checked our rifle and handgun each day along with other essential field gear, and we searched our immediate surroundings for polar bears. We were well beyond the usual hunting range of bears; however, it was important to be vigilant, as any bears found so far inland in the peak of summer would be there out of desperation and hunger and might contemplate a light snack of unwary scientists. Climate-driven changes in sea ice extent around Greenland and in other polar environments will continue to exert pressure on polar bears' hunting grounds and their biogeographic range.

The Triassic and Jurassic rock exposures (or cliffs, as you may prefer to think of them) were about a one-hour hike northwest from camp down into the valley below the plateau, at an elevation of about 258 meters. We traversed the scree-covered valley sides on hands and feet and, in places, bums. The first few days of fieldwork were taken up with prospecting for good field sites that were relatively accessible to all, contained good exposures of Triassic and Jurassic rocks, and, most importantly, yielded visible fragments—or even better, whole fossil plant parts such as leaves and branches. On our first field day, we located the boundary between our target rocks, the Kap Stewart Group, and marine sediments of the Neill Klinter Group, which rest on top. These sediments were laid down at a time when Earth's poles were completely free of ice. All the continents as we know them today were grouped into a super-continent called Pangea. The climate of Greenland was as hot and sticky as that of the Florida Keys, and the oceans around it were warm, teeming with dinner plate–sized oysters.

The important geological boundary of the Kap Stewart Group with the Neill Klinter Group was also visible, though inaccessible, from the air base at Nerlerit Inaat as a great exposure of dark Neill Klinter rocks sitting on top of the lighter—almost white—and weathered outcrops of Kap Stewart below. The juxtaposition of dark and light forms a distinctive ridge straight west of the air base known as Harris Fjeld (incidentally, named for T. M. Harris, our paleobotanical pathfinder). In terms of landscape, it represents a transition from when this part of the world was coastal land to when, through sea-level rise, it became submerged by warm, oyster-rich seas. This important transition is also reflected in the stratigraphic names *group* and *formation*. A formation is geologists' primary unit of division of rock strata. A formation can range from one to several thousand meters in thickness, but must be distinctive enough in appearance to be distinguished from surrounding rock layers, as well as thick and laterally extensive enough to plot on a map. Two or more formations may, if their strata have significant and diagnostic properties in common, be aggregated into a group, hence the Neill Klinter and Kap Stewart Groups.

The Kap Stewart sediments held within them the plant fossils through which the past landscape of this now icy region could be deciphered in all its ecological and evolutionary detail. That was our hope. However, prospecting at Ranunkeldal with hammer and chisel revealed a thick expanse of Jurassic sediments, but one that was almost completely barren of the well-preserved fossil leaves and seeds that we required as the starting point in our scientific endeavor. We combed every centimeter. Geological hammers, paint scrapers, and even a knife-edged cake server (don't ask why!) were used to cleave the finest sediment layers. After much searching, we found a few poorly preserved fragments of two fossil plants: one called *Czekanowskia*, with long, narrow leaves, and a broad-leaved conifer called *Podozamites*. They were, however, mashed up and lying in many different orientations on the sediment bedding plane, like disarticulated skeletons. Blackened fossil leaves were twisted and torn from their twigs. This suggested to us that the plant parts had traveled some distance before finding a final resting place and being buried by sediment. Although such fossils may provide some clues about distant vegetation and a distant landscape, we were specifically interested in finding fossils that showed evidence of very little transport before they were buried. These types of plant fossils are instantly stunning. There is usually an audible gasp when

FIGURE 1.3. Photos of modern landscape near Ranunkeldal.

a rock is split and they first appear out of their age-long entombment in the Earth, like the only hand-painted illustration in the pages of a long book. If you flip through a book, the illustrated pages often have a different weight of paper, which makes finding them easy. It is the same for fossil plants. If they are present in a rock sample, they form a line of weakness in the sediment. A single tap of the hammer can split the rock open to reveal the entombed fossil leaf for the first time in 200 million years. A lack of any visible damage due to transport by streams or wind, such as crumples, folds, or tears, is usually a good sign that fossils were preserved locally. These fossils represent plant species from the living vegetation that once stood close by, or in the exact location where the fossils were found. Only the voices of very locally preserved fossil specimens such as these could be combined to produce our landscape symphony of Greenland's hidden past.

Disappointingly, the cliffs that we clambered down from our plateau camp also appeared not to preserve any obvious Triassic sediments. The Triassic-Jurassic boundary marks the third greatest extinction event in Earth history, according to marine paleontologists. In the oceans, corals and reef-building organisms suffered heavy losses to extinction, and many groups of ammonites—conodonts and brachiopods, all vitally important components of marine ecosystems at the time—did not survive this transition. Similarly on land, nearly a quarter of the diverse vertebrate families that had flourished throughout much of the Triassic went extinct. We wanted to track the responses of vegetation to this strong punctuation mark in the evolutionary

record of animals, but we soon realized that this plan would not be possible at our Ranunkeldal site. We took a photo of our expedition party at the foot of the Jurassic cliffs, at the exact location of Finn's 1968 campsite, for posterity and moved our reconnaissance efforts due west to our next field location, in the heart of Buttercup Valley. We were too late in the season for buttercups, but Greenland's national flower, dwarf fireweed (*Chamerion latifolium*), was flowering in profusion. Neither of our Ranunkeldal sites had been explored by Tom Harris's original expedition parties of the 1920s and '30s, so in paleobotanical terms, we were forging new ground.

The next field site was much more rewarding. On July 20, 2002, we made our first significant discovery: large tree trunks preserved in red sandstone. The trees were probably conifers or, in the case of those with obvious side branches, more likely ginkgos. As we progress through this book, the ginkgos and their probable relatives will become familiar. Today, these plants are represented by just one species, the maidenhair tree *Ginkgo biloba*, but in the latest Triassic and earliest Jurassic, the ginkgos were a diverse group. In addition to *Ginkgo*, we will encounter *Sphenobaiera*, *Baiera*, and *Ginkgoites*. *Ginkgoites* is worthy of particular mention as, out of an abundance of caution, this scientific name is used for ancient fossil leaves that closely resemble modern *Ginkgo*, but which cannot be "proved beyond all reasonable doubt" to be such. For this reason, *Ginkgo* and *Ginkgoites* are often used interchangeably.

Returning to the red sandstone and its tree trunks, geologists would call this layer of fossil plants a *bed*, a term reserved for the smallest unit of sediment that is clearly distinguished from layers above and below it by well-defined dividing planes. This bed could be traced laterally for about 40 meters to reveal large, round cross-sections of further conifer trunks. It also contained an extinct fossil called *Neocalamites*, a giant water-loving species related to modern horsetails (in the order Equisetales). These plants, living and extinct, all have jointed stems like those of modern bamboo and distinct stem ridges and grooves running parallel to the trunk. No other species were found in this fossiliferous layer, suggesting one of two possibilities: first, that we were tracing the margin of an ancient Jurassic lake that supported a thriving but species-poor flora; or, second, that we had discovered just a fraction of the once-living vegetation's true diversity. The fact that the fossil discoveries we made in this sandstone bed were all large, woody structures, with no leaves or delicate parts preserved, suggested the second scenario. Preservation of only the toughest fossil plant parts, such as woody stems, usually indicates that conditions for fossilization were not ideal for all species. Only the most robust plant parts and species can withstand long-distance or high-energy transport by streams, rivers, or wind. There is a whole branch of paleobotany, called taphonomy, that focuses exclusively on the filtering that takes place during the transformation from a living plant community to a "dead plant community" or fossil assemblage. It is an important area of scientific study because it enables the paleobotanist to ask the question: How representative is my fossil collection of the vegetation that once lived in this place many millions of years in the past?

On July 21, 2002, our third day of fieldwork at the Ranunkeldal site, we had a clear purpose as a team. Our aim was to examine a profile extending the whole 70 meters of cliff face that we determined to belong to the Kap Stewart Group. It made up much of the north-facing side of the valley, starting from the oldest sediments at the base and rising to the youngest sediments at the top. We worked like mountain goats, zigzagging upward through this great tower of sediments and geological time, recording any and all possible evidence that would reveal what this landscape looked like in the past. We roughly estimated that every large step upward through the rocks of the cliff represented around 8,000 years in time! This is only a guess, however, which relies on there being a simple relationship between the depth of the sediment and time (which there isn't) and requires that the entire first stage of the Jurassic—the Hettangian, which dates from roughly 201 to 199 million years ago—be completely represented by the sediments of the valley side. The latter assumption remains to be tested to this day.

The 70 meters of our study section were combed over, centimeter by centimeter, for well-preserved fossil plant material. The sequence of sedimentary layers, with their physical characteristics and vertical dimensions (usually referred to as lithostratigraphy), was logged in field notebooks, and samples were placed in sturdy plastic sample bags. The logging was undertaken by Stephen Hesselbo and Finn Surlyk, both sedimentologists, from the University of Oxford and University of Copenhagen, respectively. Both are well-seasoned field geologists with the ability to hike through difficult terrain with deceptive ease. There was a good-natured but unspoken speed-hiking competition between the two, which kept the full team moving and working at a productive pace. They recorded the grain size of the sediments making up the rock types—sandstones, shales, clays. The presence or absence and types of biological activities (e.g., burrowing, walking, resting) recorded in the sediments were also noted, as these traces, much like footprints in wet sand, represented the behavior of the creatures that once dwelt in or passed through this setting. The presence of fossil roots, leaves, twigs, and wood was recorded for every centimeter of solid rock that we cracked open with a rock hammer—all while trying to avoid the daily onslaught of hairy Arctic mosquitoes. For Marlene, these details would be essential. Her role as an artist and scientific illustrator would be to reconstruct each individual plant we unearthed from these Greenland rocks. Ultimately, the individual species would be assembled into an accurate living landscape. How could we know what their ecology was like? All fossils are embedded in a matrix of surrounding rock. Steve studied the matrix of each fossil to determine whether it was sand, soil, mud, or gravel and whether it had been wet or dry. Had the plants been living in or near water, or in a dry environment? Which of them would have been grouped together, in association? Association is an important element in any ecosystem, as it describes the relationships between the plants and their environment.

We extricated fossils from their rocky matrix with the smash of hammers. Six days of hard fieldwork had yielded hundreds of double-bagged rock samples for further lab analysis, over 100 collective pages of field notes and detailed field logs on the sequence of rock types that made up the cliffs' overall structure. Lithostratigraphy follows similar principles to the judging of a layered cake in a bake-off competition, where the depth of the bottom sponge layer is measured and characterized (texture, firmness, color—it's ill-advised to taste too many rocks, though some geologists have been known to distinguish silt from clay by seeing if a sample feels gritty on the teeth when chewed, in which case it's silt!), followed by the mousse layer, then another sponge layer, and the fondant icing layer on top. This cake stratigraphy may record fruit or chocolate chips in certain layers. Similarly, we recorded the rock stratigraphy in field logs for future reference and noted all layers where fossil plants occurred. Views from our lunchtime rendezvous point confirmed that the majority of the 115-meter cliff consisted of sandstone and shale layers. These layers were formed over 200 million years ago, their sediments deposited by high-energy rivers flowing into the freshwater Kap Stewart Lake. We determined from the fossils and their associated sediment matrix that the lake was fringed by wetlands dominated by the towering bamboo-like *Neocalamites* in a balmy humid and hot climate.

Neither shale nor sandstone is ideal for the preservation of fossil leaves. Plant fossils are best formed in acidic or oxygen-poor environments where microbial decomposition of plant parts is halted completely, or at least slowed down. When an undecomposed leaf with a thin microbial film is covered in sediment of some sort, and then lightly baked and compressed as the sediments are turned to rock (lithified) over millions of years, a fossil leaf is formed. These leaves don't remain green, but darken toward black, as their chemical composition is altered and they begin to turn to coal. Under the right conditions, these fossilized leaves can retain their fine cellular detail in the waxy, polymer-rich substance, called the cuticle, that covers the entire leaf surface. Microscopic analysis of these "coalified compression fossils" reveals stomata, tiny breathing pores that breach the impermeable cuticle, as well as delicate glands, hairs, and cell wall outlines of the leaf "skin," which we call the epidermis. Although Ranunkeldal did not yield the bounty of fossil plants we had hoped for, it provided an important demarcation for the northernmost extent of the ancient Kap Stewart Lake. It also indicated that this great ancient lake was fringed by marshy thickets of giant *Neocalamites* in Jurassic times, grading into woodlands with stout-trunked conifers that bore broad rather than needle leaves, and feathery-leaved *Czekanowskia*, a probable *Ginkgo* relative. Our team was pleased and excited by the finds at Ranunkeldal. We were primed and ready to move onto the next expedition site. What was in store for us at Astartekløft?

ASTARTEKLØFT

Our party was to arrive by helicopter at Astartekløft on July 27, 2002, down by one. We had lost Finn, certainly the person in our party most familiar with the geology of Greenland, but also the one who had the most experience navigating and surviving in this harsh environment. Finn's departure from the group came in dramatic fashion back in Ranunkeldal, where, as we scrambled across a rough river gorge (using hands as much as feet), Finn had slipped on an algae-covered river boulder and fallen into sharp jutting rocks, breaking his thumb with a compound fracture. We had to make a quick decision about the best evacuation options. Roughly bandaged, bleeding, and with bone protruding, Finn was in no fit state to climb the steep valley sides for a helicopter rescue, yet fog was beginning to move into the valley. Worse, we had no satellite phone service on the valley floor, and Finn, sinking into shock, was the one person in our party who was trained to administer morphine and guide a helicopter landing. Two of us gave him basic first aid, while one scrambled to the top of the valley to call in a helicopter rescue from Nerlerit Inaat, a 40-minute flight away. Despite his significant injury, it was Finn who guided the helicopter to a safe landing, one-armed, in his orange and silver emergency blanket.

Despite this inauspicious start, as we explored the strata at Astartekløft, the opportunities they offered for in-depth understanding of the biotic events that occurred during the Triassic-Jurassic mass extinction rapidly became apparent. In fact, it was soon clear that, in the words of our colleague from the University of Bucharest, Dr Mihai Popa, we had "hit paleobotanical gold." Astartekløft was the site most likely to provide us with the fossil data we needed to understand the epic changes that Greenland's ancient ecosystems underwent during this period of catastrophic climate change. This incredible fossil-rich locality got its name, Astartekløft, from a fossil bivalve (clam) belonging to the genus *Astarte*, which gave its name to the River Astarte, and the Danish word for cleft or split (*kløft*). This cleft that cut into the Kap Stewart cliffs made accessible (accessible, that is, if you are either a nimble Arctic fox or a determined geologist prepared to scramble up a precipitous gully) layers of rock that spanned the extinction event, our target for the expedition. This site was to become our equivalent of the Rosetta Stone, allowing us to translate a lost language into a modern understanding, and its rediscovery marked the onset of a long collaboration between scientist and artist in an attempt to both explain and make accessible to all one of the most profound climate-driven extinction events the world has known. The fossils of Astartekløft allowed us to piece together, bed by bed, the floral history of this part of Greenland during the latest Triassic and earliest Jurassic, ultimately revealing that about 200 million years ago, what is now the Arctic was then vegetated by a lush, almost tropical flora. Welcome to Astartekløft, our Tropical Arctic.

Within 100 meters of our Astarte River field camp, we discovered fossil plant–laden rocks—so many, in fact, that almost every piece of scree at the base of the cliffs was imprinted with rust-red and blackened leaf fossils. We then had to work out exactly which layers or beds each fossil had fallen from to make sense of the passage of geological time and the shifting of landscapes that these cliffs recorded. Over the course of a fortnight, we pieced the evolutionary story together. At the base of the cliff was the Horsetail Bed (Bed 1), rich in the remains of ridged and jointed stems of *Equisetites* (fossil horsetails). We had seen a larger related species (*Neocalamites*) in the marshes that fringed Kap Stewart Lake at Ranunkeldal. The Halfway Bed (Bed 1.5) followed a little farther up the cliff, with sediments suggesting a halfway point between totally dry terrestrial and sodden wet floodplain environments. The Halfway Bed was a serendipitous find on the penultimate day of fieldwork. Next was the *Ginkgo* Bed (Bed 2), which preserved thick fossilized leaf mats of beautiful *Ginkgoites*. Our 200 million-year-old leaves from the *Ginkgo* Bed, sandwiched on top of one another, evoked for us fall in a wild ginkgo forest, a rare sight that now can be witnessed in only a few places in the world, such as the Dalou Mountains of southwest China. A few meters farther up the cliff face was what soon became known to us as the Fox Bed (Bed 3). This bed marked the level in the cliffs where we sat helplessly one morning watching an Arctic fox raid our kitchen stores in camp. Carefully, she pried into the food trunks, then trotted off some distance with her stash before burying it. The Fox Bed was rich in fossil leaves and stems of the conifer *Podozamites*, with its broad leaves reminiscent of modern kauri pine from New Zealand and Australia. The Flower Bed (Bed 4) was yet farther up the cliff and away from camp. It was chock-full of hundreds of tough, wrinkled fossil bracts (the outer protective parts of flowerlike structures) called *Cycadolepis*, belonging to a long-extinct group called Bennettitales. Was this a sign of impending ecological doom? Mass flowering events in modern times, such as desert blooms following a 100-year rain event, can signal stress or rare opportunity. Somewhat ominously, the Flower Bed was succeeded by the Boundary Bed (Bed 5) a few meters up. This horizon marked the boundary between the Triassic and Jurassic periods and a mass extinction of global fauna. Collectively, these six beds of fossils recorded snapshots in time at the close of the great Triassic period.

The next set of fossil beds were all of Jurassic age. The Disaster Bed (Bed 6) was the first-time slice preserved in the immediate aftermath of the Triassic-Jurassic extinction. It was going to be a critical window into a very changed world, where volcanic eruptions were at their peak and climate warming had created a global hothouse. Would it tell us what the landscape of Greenland looked like during such a time of environmental crisis? Finally, near the top of the cliff face, two farther beds, the *Spectabilis* Bed (Bed 7), named after the spectacularly large-leaved tree *Sphenobaiera spectabilis* that enriched its siltstone layers, and the Ledge Bed (Bed 8), with the best views of the valley, completed the picture. Both provided snapshots of Jurassic time in which we hoped and expected to see some sort of ecological recovery following the environmental crisis and extinctions encapsulated in the Boundary and Disaster Beds. We pondered what they would tell us about life and landscapes in the aftermath of a major mass extinction event.

Our search for and rediscovery of these elusive fossil plant beds was not a needle-in-the-haystack affair. Fossil plants at Astartekløft owe their original discovery to a nineteenth-century Yorkshire whaler and clergyman, William Scoresby Jr. He was the first to map what would later come to be recognized as the world's largest fjord, Scoresby Sound. While mapping, Scoresby recognized plant fossils, though it was not until 1891–92 that they were collected, first by Danish geologist Nikolaj Hartz at Kap Stewart in Jameson Land, and then later in the 1920s and '30s at Astartekløft by University of Reading professor Tom Harris.

We had targeted our approach from the start using the original field notes Harris had scribbled 80 years earlier in the exact same locality. In fact, he probably camped where we did, nestled down at the base of the valley by the stream. His field notes allowed us to locate each of Harris's original plant beds almost to the exact stratigraphic level. We rediscovered his *Lepidopteris* Bed (named after a plant species that went extinct at the Triassic-Jurassic boundary), containing great, almost fleshy chunks of blackened leaves sandwiched in ancient leaf litter, at 10.9 meters above the valley floor, then worked our way up the cliffs, in mountain-goat style once again, until we found Harris's last bed at 80.9 meters. As with all scientific endeavors, it was critical that we were able to collect the fossil plants of East Greenland in the context of previous seminal work.

The rocks at Astartekløft contrasted sharply with those that we had clambered up and down every day at Ranunkeldal. Although they were the same age, those at Astartekløft comprised sand-rich sediments, denoting rivers, that were regularly interspersed with layers rich in silt and clay and packed full of fossilized plants. These finer sediment layers were laid down when the deep rivers spilled over their banks and deposited their heavy load of fine silt and clay sediments across the floodplains. The discrete silt- and clay-rich interdigitations that marked the locations of our fossil plant beds ranged in thickness from a few centimeters to over 2 meters, suggesting that the river systems of this ancient delta environment were a powerful force in shaping the landscape. According to the sediment data, they regularly washed the floodplains with mineral-rich sediments, perhaps similar to those deposited by the whitewater rivers of the Amazon Basin today. The Solimões and Japurá Rivers in modern Brazil, for example, release up to 20 centimeters of their fine sediment load onto their floodplains annually, and the regional várzea forests have to withstand inundation by floodwaters for up to a third of the year. Plenty of sediment, together with mildly anoxic floodplain conditions, is a prerequisite for excellent fossil plant preservation. These conditions act to arrest bacterial and fungal breakdown, and thus leaves, seeds, and twigs of the forest floor become entombed and fossilized by the river sediments that are spilled on top of them.

On July 29, once we had located and recorded all the fossil plant beds we had journeyed to Greenland to explore, we started fossil hunting in earnest. We used a search method that previous researchers had designed to elucidate changes in landscapes and floras through time. This was the first time that these rich Greenland sediments would be scrutinized in such a fashion. In the 1920s, Tom Harris collected a trove of fossil plants from the same location at Astartekløft to document their ancient biodiversity. However, he did not conduct a fossil hunt with ecology in mind. Without ecological understanding, one cannot assess whether the species that went extinct in a place and time were the dominant forest trees that covered entire landscapes or were rare and elusive specialists hiding out in a single location. Our aim was to add this important layer of understanding. We wanted to reconstruct ancient Greenland as a real living ecological landscape, not just a checklist of fossil plant species that were present in some beds but then disappeared. To achieve this lofty aim, our fossil hunting recipe was designed to minimize the potential biases that could muddy our interpretation of a living landscape from the few long-dead fossil fragments left behind. The potential biases are, unfortunately, numerous. For example, a personal bias toward discovering new fossil plant species never before described by any other scientist would have colored our interpretation of the once-living landscape as teeming with plant species that in reality were quite rare. We did not want to approach the fossils like a modern bird-watcher whose species checklist records both the house sparrow and the California condor with equal weight—both get a tick on the list, yet the sparrow is very common and one tick may represent many sightings, while the condor is rare and the tick here may represent just one fleeting glimpse. Our method of fossil collecting therefore added a layer to Harris's checklist approach. We estimated how rare or common the ancient plant species of East Greenland were by counting the number of fossil leaves we found belonging to each species. In a modern forest, studies have shown that the tree species with the highest numbers of individual leaves in the litter on the forest floor are the most common or ecologically abundant at the landscape scale. Similarly, the species with the lowest numbers of leaves in the litter are usually ecologically rare.

In the long, often chilly summer evenings, usually spent in the kitchen tent, we mulled over ways to design a fair comparison of the fossil bounty unearthed from each sediment bed. Some of the beds, such as the Boundary Bed, yielded thousands of fossils, while others, such as the Halfway Bed, yielded fewer than a hundred, despite the fact that we spent the same amount of time and effort excavating each. Does this difference have ecological meaning for reconstructing past environments? Well, it certainly causes problems when you want to fathom and visualize past biodiversity. If you were designing a perfect experiment to test any hypothesis, you would never start with a sample size of 100 for one experimental treatment and a sample size of over 1,000 for the other. We can use statistics to "fix," or normalize, such inequalities in sample number. What is more complicated in paleoecology is getting your head around imperfections that are introduced into your perfect experiment, but which cannot be "fixed" with statistics alone. As Mark Twain remarked, "There are three kinds of lies: lies, damned lies, and statistics"—this was a trap we needed to avoid. In paleobotany, we rely on the fossil leaf litter to interpret the living vegetation that produced it. Ecological studies show that there are many types of locations and environments where living leaves fall and are covered with sediment. In the trade, these are called depositional environments. Each is subject to a unique set of biases that can result in distinct filtering of the living plants and forest communities they represent. Think about an autumn day and imagine wading through orange- and red-tinged leaves that have recently fallen. It would be rare to see the delicate, slender leaves of spring snowdrops among them. As a contrast, now imagine a summer storm that has whipped the leaves off *all* plants, deciduous, evergreen, and delicate herbs on the forest floor. These leaves may be more damaged and more mashed up when preserved

OY-HIE

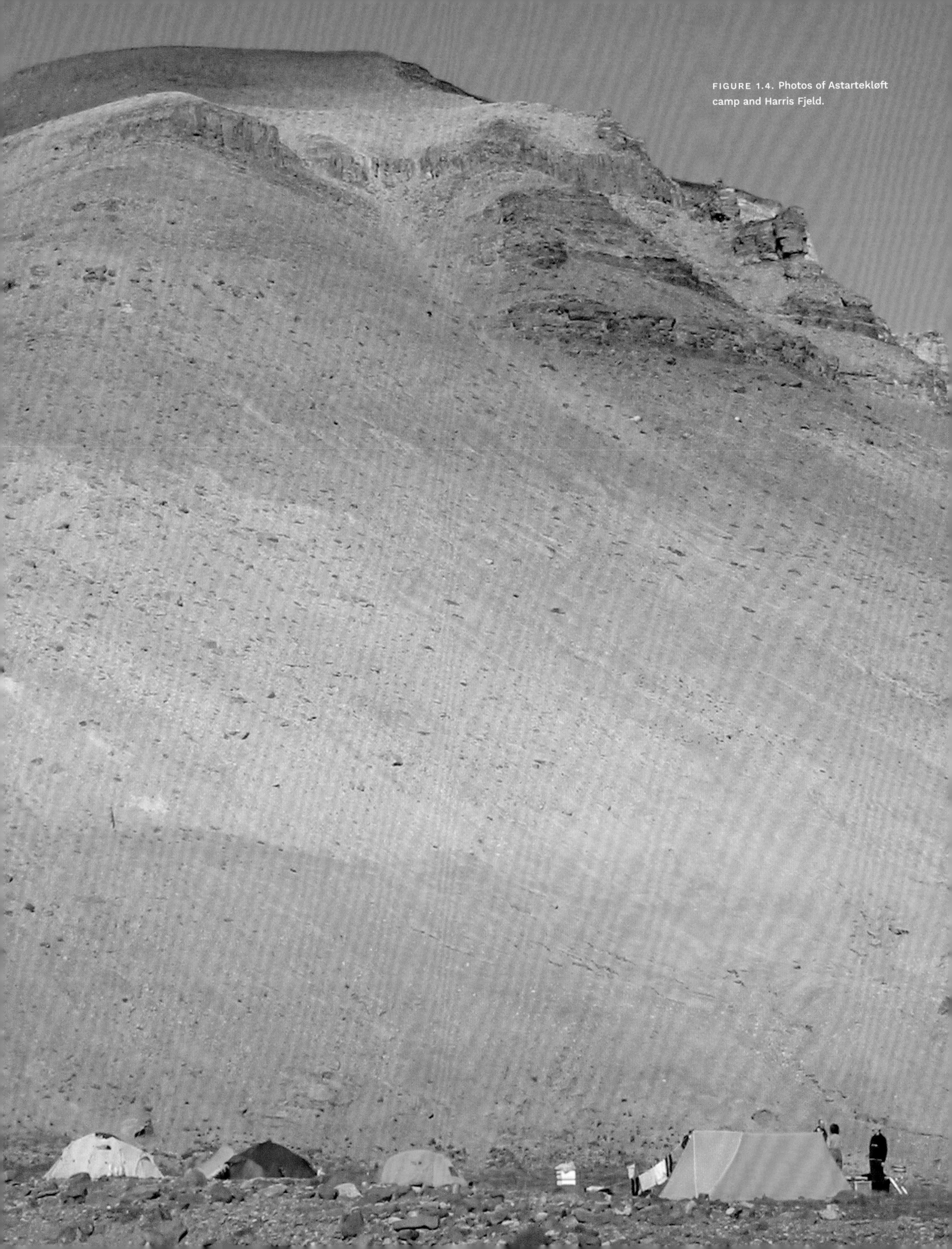

FIGURE 1.4. Photos of Astartekløft camp and Harris Fjeld.

as fossils, but they are perhaps more representative of the living forest communities from which they were stripped.

The usual way of fixing the problem of complex biases in paleobotany is to take an isotaphonomic approach—meaning that we only compare like with like. For example, ecological insights gained from the fossils formed during massive flooding events are compared only with ecological data collected from similar deposits or similar fossilization pathways. Luckily for us, the same type of flooding event formed all six Triassic fossil plant beds at Astartekløft, thus providing detailed portal views of the last few million years of the Triassic, when coral reefs were beginning to collapse and terrestrial ecosystems were poised for massive environmental upheaval driven by intense volcanism associated with the opening up of the proto-Atlantic Ocean. As we moved through geological time with each boot step upward through the Astartekløft cliff sediments, we were assured, therefore, that any obvious change in the number of fossil leaves recorded for each species was likely to represent a real signal that the ecology of the living landscape was changing—that the composition of the forests was fluctuating.

Our fossil hunting method was a hybrid between paleobotanical and modern ecological approaches. Every morning, following breakfast, we hiked the very short distance from camp to the base of the cliff section. Each of the four fieldworkers was tasked with a fossil bed excavation that lasted for precisely the same duration, to avoid introducing collector bias. We excavated at distant points along the same bed to trace it laterally across the paleo-landscape. By doing so, we would later be able to assess whether the forested ecosystems of the time were varied and patchy or uniform across the landscape. In practice, this involved digging oneself into a small ledge at the top of a plant bed and excavating roughly 1 cubic meter of sediment downward with a hammer and chisel each day. This strategy was meant to ensure that every fossil plant bed was represented by an equivalent volume of rock and was sampled for the same amount of time. We wrapped each fossil discovered, without exception, and labeled it with a unique sample number, provisional ID, date, and collector's name. These samples formed the database from which the diversity and ecology of the ancient forest systems of this place and time would be reconstructed. If there had been any spectators, we would have been an intriguing sight: four scientists with mosquito head nets, hammers, chisels, toilet paper (for wrapping!), masking tape, and stacks of Danish and Icelandic newspapers (for wrapping too), hunkered down on the cliffs of the Astarte River valley. The fact that we were all sitting on top of an ancient delta floodplain, surrounded in our little excavation pits by leaves and twigs that had once belonged to extinct species from 200 million-year-old woodlands, highlighted the absolute wonder of paleobotany.

It took us six days to excavate our way through Rhaetian time—all 7.2 million years of it. Starting with the Horsetail Bed at the bottom of the cliffs, we spent about a day excavating each of the six Triassic fossil plant beds. We reached the Triassic-Jurassic boundary (Boundary Bed) on August 2, 2002, with a lot of excitement and many questions. We felt like Tenzing Norgay and Edmund Hillary cresting Everest. Many years have gone by since our 2002 expedition, yet the nature of paleobotany and the careful methods we employed mean that the information we extracted from the Triassic rocks of Astartekløft is unchanged. Today, it is more relevant than ever. Although human-caused climate change and its impacts have accelerated in Greenland and all over the world in the past two decades, modern climate change cannot alter or erase the paleobotanical story locked up in the Astartekløft rocks. When glaciers and ice sheets melt, as they are doing in Greenland today, the bubbles of ancient air locked in old ice are released, permanently obliterating these long-term records of atmospheric change. However, on a human time scale, it would take an absolutely catastrophic event on the order of a massive meteorite impact to obliterate forever the miles and miles of rock cliffs that crop out along the shores of Jameson Land and the fossils they contain. Excepting geological processes, only this type of event would destroy forever the deep-time geological record of Greenland's ancient climate, landscape, flora, and fauna and the message they hold for modern humanity.

Bed 5, our Triassic-Jurassic Boundary Bed, remains the most famous of the beds at the Astartekløft field site. It records a snapshot in geological time when mass extinction was in full swing. Species lost in the ocean included bivalves and brachiopods as well as coral reef–building biota, including sponges and corals. On land, phytosaurs, crocodile lookalikes that were some of the largest land-based predators; the beaked bipedal dinosaur-like Shuvosauridae; and reptilian aetosaurs, named for their supposed eagle-like skulls and lizard-like bodies, were all suffering reductions in their population size. All three groups, plus proto-dinosaur groups like the small-bodied Silesauridae, went extinct at the Triassic-Jurassic boundary. In total, nearly a quarter of Late Triassic animal families went extinct globally, which is a loss of biodiversity on a remarkable scale. We wanted to know whether plant life suffered the same levels of extinction as animals. Tom Harris's pioneering research back in the 1920s and '30s showed without doubt that 85 percent of the fossil plant species found in the Late Triassic beds of Greenland suddenly disappeared around the time represented by our Boundary Bed. We wanted to delve into these data in more detail. Did these sudden disappearances represent the extinction of plant species? During our excavation of the Boundary Bed, ecological questions were constantly in the back of our minds. What about the 15 percent of fossil species that did appear in Jurassic rocks—were these the survivors? How did they survive—what characteristics did they possess that enabled them to survive at a time when so many animal families, and up to 80 percent of marine invertebrate species, went extinct globally? Watching helplessly from the Fox Bed as our food rations in the valley below were repeatedly raided by a determined Arctic fox threw up more than a few hints as to those characteristics. Our research now had to pinpoint those characteristics in ancient fossilized plants, as they might be relevant to our modern situation and changing climate. What a challenge!

One cannot answer all research questions in the field. The purpose of our fieldwork was to establish biological and environmental associations—for example, what plant species co-occurred, and what ancient environments did they occupy? The number crunching and the detection of ecological trends or extinction events was to take place back in the lab. Striking glimpses of an ancient biodiversity crisis were, however, immediately obvious as we excavated our way through the sandstone and mudstone layers of the Boundary Bed. For example, a fern species, *Thaumatopteris brauniana*, which we had not encountered in the preceding five Triassic beds lower down the cliff section, appeared for the first time. This little fern fossil is related to modern ferns within the Dipteridaceae, which occur today only in tropical climates. It marked the transition from the Triassic to the Jurassic around the top of the Boundary Bed. We were nearly in the Jurassic—the Age of Dinosaurs!

By August 5—the date when we exhumed the first fossil plants of undoubtedly Jurassic age—we knew that we were pushing our luck in terms of the Arctic weather. The average monthly temperature for July in the region of Astartekløft today is 3.3°C (37.9°F). Temperatures fall frigidly to an average of −0.4°C (31°F) by September, and snow can start any day in August. The sunniness of this place had taken many of us by surprise, however. In fact, for the months of June and July, Astartekløft has an average number of sunshine hours (200–250) comparable to that of Miami, Florida, although its air temperature is some 25°C (45°F) cooler. All of the fossils collected from our study sites thus far hinted at a much warmer prevailing climate for Greenland 200 million years ago—more akin to Miami and the Florida Everglades than the polar desert of modern Astartekløft. The scientific basis for such a hint would, however, have to wait for proper testing back in the lab.

Our first unquestionably Jurassic fossils were unearthed from the Disaster Bed. This thin little wedge of a fossil plant bed was immediately interesting, but troublesome because of the amount of black, organic-rich decayed plant matter it contained—not enough for it to be classed as a coal, which would indicate an ancient peatland, but too much for it to be a riverbank deposit or part of the floodplain. What did the landscape look like at this time and place? It couldn't have been a swamp, as there were too few fossils of tree species; rather, it was dominated by ferns, such as *Cladophlebis* and *Thaumatopteris*. We concluded that it must have been a sodden wetland in which an abundance of decayed ferns combined with inorganic mud to form a thick, sticky ooze, all in the aftermath of a global mass extinction event. This atypical bed would be pivotal to our understanding of the impact on the flora of the volcanism and climate warming that were in train at that time. All the fossils found in each of our excavation pits within the Disaster Bed were collected and wrapped for further analysis, following the methodology we applied to all the other fossiliferous layers at the Greenland sites. We were acutely aware, however, that any vegetation changes detected between the Disaster Bed and the preceding Boundary Bed would have to take into account the fact that the Disaster Bed was not a flood deposit like all the other Triassic fossil plant beds we had discovered. The Disaster Bed preserved very local plant remains from a Jurassic marsh-like flora, whereas the preceding six beds encased leaves, twigs, and cones from flooding Triassic rivers. This distinction may seem overly semantic, but from a standpoint of academic rigor, it was vital information. Our task was to tease apart real ecological change, using evidence from the fossil leaves we had excavated, and false signals of ecological change. False signals of an ancient biodiversity crisis can be caused by differences in how fossils are made in different environments. For example, swampy and marshy environments tend to preserve only locally growing plants, whereas floodplains can preserve and fossilize not only local plants, but also those brought in by powerful floodwaters from some considerable distance away. False signals can mask real ecological change or, worse still, can suggest ecological change when, in reality, none has occurred. Therefore, rather ironically, one of the challenges of paleobotany is "seeing the wood for the trees."

We encountered our first portal into an ancient oxbow lake from Jurassic times on August 5. Just as architects can visualize the three dimensions of a room from drawings, a sedimentologist can reconstruct historical and even truly ancient environmental settings by carefully observing the characteristics and organization of sediment grains in a stack of rock layers. For sedimentologist Stephen Hesselbo, one of the telltale signs that the *Spectabilis* Bed was an ancient oxbow lake was the fact that the sediments were coarser at the bottom of the bed and became finer and finer toward the top. This subtle sedimentary signature told the story of a braided stream or river meander that, around 199 million years ago, finally separated part of itself from the main stream to form a little lake. Finer and finer sediments would have been deposited as the water flow quieted from the fast current of a river to the cauterized stillness of an oxbow lake. The plant fossils in this bed were absolutely exquisite. Huge hand-shaped fossil leaves of *Sphenobaiera spectabilis* packed the bed. The species name, given by Swedish explorer and geologist Alfred Gabriel Nathorst, acknowledges the stunning appearance of these fossils: *spectabilis* means "admirable" or "spectacular." The same species was still present in the Ledge Bed a little higher up the cliff face, or "up section" as we would say in geology, yet not quite as exquisitely preserved. Long, lacy leaves of the fossil plant *Czekanowskia* were also abundant in the Ledge Bed, as they had been in the *Spectabilis* Bed below and in Jurassic deposits over 40 minutes' helicopter ride away at Ranunkeldal. The fossil leaves of these last two Astartekløft beds, our portal into the Jurassic world, were notably thick in the field and lifted off the rock surface like a modern leaf litter. A single chisel strike of the rock they were encased within opened them to the air. Whole broken chunks of 199 million-year-old fossil leaves lifted up and fluttered if the wind was strong enough. Even their original venation pattern was visible if you held them up to the watery Arctic sunlight.

Our fieldwork at Astartekløft camp yielded just over a metric ton of fossil plant and rock samples for detailed analysis—a veritable bounty. The samples were wrapped and boxed up into "fossil depots" at our campsite for the first leg of their journey, by helicopter back to the air base at Nerlerit Inaat. They were then loaded as freight onto a bright red Royal Arctic Line ship heading to Aalborg, Denmark's fourth largest city. This precious cargo, which represented our 200 million-year-old time capsule, was then shipped across the Atlantic, past Newfoundland, and down the Gulf of St. Lawrence to the Great Lakes port of Cleveland, Ohio. From there, the samples traveled by container truck to the Field Museum of Natural History in Chicago, where two scientists and one artist waited nervously by the loading dock. We watched as the forklift raised our window onto the Late Triassic and Early Jurassic world of Greenland into the summer warmth of the museum's loading bay.

BACK IN THE LAB

Such was the weight of the fossils that we had discovered and dug from our little excavation pits dotted up and down the valley side of Astartekløft that our astronomical helicopter bill would, in time, become a thing of Field Museum legend. However, to us, this was a welcome expenditure, signifying collecting success. The arrival of the fossils at the Field Museum earmarked the next phase of our study. Now, we wanted to put these precious fossils to use, to begin our reconstruction of Greenland's ancient landscape and to start to unravel their story. As they were unloaded, we hoped that these fossils would soon start to fulfill the promise we felt they offered while unearthing them in the field. Could they enable us to track the all-important extinction event in the form of landscape and vegetation changes, bed by bed, through this 10 million-year time interval in Earth's history? Could we fuse the practices of science and art to portray the events that transpired in Greenland from the fossils of the ancient Kap Stewart Lake region?

Weeks upon weeks were initially spent unwrapping, labeling, identifying, and curating the fossil bounty, a task that occupied paleobotanists Popa, Glasspool, and McElwain and a small army of industrious museum volunteers. In paleobotany, we determine the species of a fossil by its similarity to other fossils. We do not require strict adherence to the modern biological species concept, which states that the individuals of a species must be capable of interbreeding. This is simply because we cannot reliably observe the mating behavior of fossil species! Identification of a species in paleobotany is therefore usually based on a morphological species concept, or, using less jargon, a species that is defined by its size, shape, and structural characteristics alone. These characteristics include microscopic traits, such as leaf hairs and stomatal features, that are visible only with high-powered microscopes. We were acutely conscious, however, that the time period we were studying was one of profound environmental change, and that plants have a habit of adapting their morphology and anatomy to the prevailing climate. This propensity can confuse or blur the identification of species. To avoid this potential confusion, we identified the majority of our fossil finds to genus rather than species. Were we anthropologists studying human evolution, this approach would be equivalent to identifying the ancient skulls of hominids with many similarities to us (*Homo sapiens*) only to the genus *Homo*, rather than differentiating them from us and identifying them to species, such as *Homo erectus* or *Homo antecessor*.

Once our preliminary survey was completed, we had identified and catalogued about 4,200 fossil leaves and 122 seeds, cones, and flower parts. Our efforts resulted in the first ecological database of Triassic and Jurassic plants from East Greenland, providing vital information on which species were rare and which were abundant and how each species trended through time. What next? Ultimately, we were looking for the ecological traits of each fossil that were equivalent to those of the wily Arctic fox who stole and buried our dinner supplies. We were searching for species that survived in the midst of a mass extinction event and the features of those species that allowed them to survive. The database was thus used to characterize the ancient flora. It was used to identify the fossil species present or absent at each instant in time (represented by a fossil plant bed) and to estimate the local plant biodiversity at that time. We calculated the relative abundances of the fossil species present in order to identify which species were the most abundant, and thus ecologically dominant in the landscape of the past, and which were low in abundance, and therefore presumed to have been ecologically rare. Finally, because we had so diligently excavated at many different points along each of the fossil beds in our personal excavation pits, we were also able to assess the heterogeneity of plant communities across the landscape. This was vital information on association that Marlene needed to join all the words into a language, or, rather, assign all the plant species to their right place in the landscape.

We investigated the fossil plant database using complex statistical analyses. These tests enabled us to take into account some of the biases prevalent when working with fossils. An obvious bias is caused by differences in the number of fossils excavated from each bed. For example, our fossil hunting endeavors yielded a whopping 1,023 fossils from the Boundary Bed, spanning the time of transition from the Triassic to the Jurassic, but only 62 from the older Halfway Bed. Diversity was a key component for the reconstruction of Greenland's past landscape; however, to assess it ecologically in its broadest form, we did not want to simply estimate how many plant species were present. We also wanted to try to estimate how these ancient species shared the landscape with other species. Were some species space hogs that dominated every available suitable niche, or did all species share the "ecological space" evenly? Highly diverse and productive modern ecosystems such as tropical forests are rich in species numbers, but they also have high evenness values, which means that all the species present share available niches relatively equally, rather than some species dominating and "bullying" all others. Evenness at Astartekløft was measured by calculating the relative abundance of each fossil species in each fossil plant bed. These data were then visualized using pie charts for each fossil plant bed to assess how big or small each species' slice of the total ecological pie was.

For the scientists, the landscape of East Greenland was coming into focus. We had a good idea of its historical biodiversity from our estimates of the number of species in each bed (we refer to this as plant species richness), how even the distribution of species across the landscape was (we call this metric evenness, and we find that the biodiversity hotspots in today's world are also the most even), and how thoroughly the forest species were mixed at a landscape scale (a measure of spatial heterogeneity). Richness, evenness,

FIGURE 1.5. Reconstructing *Podozamites* from fossils through modeling, sketching, and computer rendering.

and heterogeneity—all are key components of biodiversity. The higher these measures are for the plants at the base of the food chain, the higher the overall floral *and* faunal biodiversity is.

These measures of biodiversity are the bread-and-butter concepts of the ecologist. Familiarity with such concepts and their numerical estimation allowed us as scientists to place what appear to be abstract numbers into a modern ecological context. Frankly, however, these terms and values were not the best means of communicating the lost landscapes we were painstakingly piecing together. A careful translation to the visual was urgently required. We handed over pie charts of species abundance, diversity data, and species names for each fossil to Marlene in a set of colored folders. We had discovered a plant biodiversity crash of epic proportions that started with early evidence for ecological instability in the Flower Bed, followed by a complete ecosystem collapse in the Boundary Bed, but we wouldn't know the full extent of what we had left to learn about the changing nature of these Greenland landscapes until we asked a different set of questions.

Marlene wanted to *see* the fossils and their landscape context, not so much the numbers. She wanted to know which fossils from the thousands of pounds of specimens embedded in rock matrix belonged together, and which were part of the same plant, and how we knew. We selected the best fossil specimens for Marlene and showed her what to look out for. She carefully drew each fossil species while looking through a microscope, fitting everything together, with the scientists checking every step. Fossil leaves that

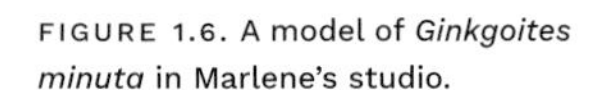

FIGURE 1.6. A model of *Ginkgoites minuta* in Marlene's studio.

FIGURE 1.7. Marlene Hill Donnelly, *Ginkgo biloba study.* Watercolor and pencil.

have laid in repose under a mountain for 200 million years or so tend to be very flat. So how to figure out how they had looked in three-dimensional life? Marlene constructed models out of paper, foil, and wire. She traced and duplicated the life-sized fossil leaves and backed them with adhesive foil. Next she cut bonsai wire for the leaf midribs and petioles (leaf stems), then attached each petiole to a larger piece of wire representing the main stem.

Both foil and wire are flexible and are easily bent and shaped to lifelike positions. Thus Marlene brought the fossil leaves to life, but did not stop there. Leaves need branches, and branches need a trunk, but these were rarely to be found in our fossil trove from Astartekløft. How did we know what leaf positions would be most accurate? First, we looked for living analogs (nearly equivalent plants); for instance, to reconstruct ancient *Ginkgoites*, we used the modern *Ginkgo biloba*, and for ancient *Podozamites*, living *Podocarpus*. "Easy" for Marlene! Bringing to life extinct plants that have absolutely no living analog or relative—that was a bigger challenge, which we will explore in later chapters on this journey.

2 Forests of a Lost Landscape

The fossils buried deepest in the Astartekløft cliffs—from Triassic times—revealed resplendent, verdant riverbank forests, whose towering trees provided homes for a variety of species in their canopies and shaded the ground below. The Horsetail, Halfway, and *Ginkgo* Beds all shared common ecological features. All told a story of biodiverse, stable forested riverbanks. These Triassic forests intermingled with more open, unforested expanses stretching far out across the river floodplain, where the plant incumbents were knee-deep in water. The fossils pick-axed from the Horsetail Bed were not all perfectly preserved. Some of the fossil leaves were a little crumpled and incomplete, and they did not all lie in the same orientation on the bedding plane. This suggested to us that a number of the Horsetail Bed fossils had traveled quite a distance as leaves and twigs, and had been broken up and damaged in the turbulence of the water, before being covered in sediment and preserved. Other fossil leaves looked like they had fallen pristinely from a tree just a few feet away from the place where they were fossilized. We inferred, therefore, that the Horsetail Bed included some species from the immediate floodplain environments and others from gallery forests lining the riverbanks farther upstream. In contrast, all fossils found in the Halfway Bed were perfect, photogenic, and oriented like soldiers in formation—all aligned. They had not traveled far on turbulent river waters; instead, the trees and shrubs on which they were growing had shed these leaves and seeds quietly at their feet.

Together, these insights from the sediments and fossil plants allowed us

FIGURE 2.1. Marlene Hill Donnelly, *Forest sketch*. Charcoal and pencil.

to argue that two vegetation types, or *ecotypes* (geographic varieties or populations within a species that are adapted to specific environmental conditions), were present at the landscape scale at this time in Greenland's deep past: a terra firma group, and a floodplain group that could withstand inundation. The subtle elevation differences between these ecotypes formed the backbone of the Late Triassic landscape reconstruction. We first imagined the landscape from the perspective of the riverbank. The view would include plant species characterizing the slightly elevated forests of the riverbank, sweeping out toward a different grouping of species occupying the floodplain. Over the next pages, we will illustrate the layers of scientific information unearthed from the fossils and the artistic and engineering endeavor undertaken to bring these 200 million-year-old fossils, and the vast landscape they occupied, to life.

LUSH

Leaf fossils mined from the oldest Triassic beds provided ample evidence that the vegetation of the time was lush, diverse, and complex, in sharp contrast with the Arctic tundra that prevails at latitudes of 70° N to 72° N today. A diversity of plant life-forms, leaf shapes, and architectures were inferred from the fossils. These provided both the skeleton and the detail to bring the Late Triassic landscape to life. Detailed measurements of leaf shape were plotted in three-dimensional space to assess how the Late Triassic forests compared with contemporary forests—were they more or less diverse? The ecological strategies of the fossil species were assessed by applying simple models that can help to distinguish between deciduous and evergreen strategies. The waviness of the outer walls of leaf skin cells (epidermal cells) was visualized using UV fluorescence microscopy to work out the likely habit of each species and whether it occupied the shady understory or the high sunlit canopy. Particularly wavy, almost jigsaw-like patterns indicate shady conditions, while straight-walled epidermal cells usually point to high-light

FIGURE 2.2. Marlene Hill Donnelly, *Triassic preliminary sketch*. Watercolor and ink.

FIGURE 2.3. Marlene Hill Donnelly, *Lush: A Portal into the Late Triassic.*

environments. Other paleoecological hints for each fossil were investigated using anatomical and chemical clues.

In the foreground of this first glimpse of the Triassic are the woody trees that bore *Czekanowskia* and *Ginkgoites* and the vine-like *Lepidopteris ottonis*. In the dark background, through a gap in the upper canopy, we see the buttress trunk of the deciduous conifer *Stachyotaxus septentrionalis* rising from the inundated floodplain. The fossil leaves of this species have scars at the ends of feather-like shoots, indicating that they were not shed as individual needles, as in modern Christmas trees. Instead, the leaves were dropped while still clinging to their little woody branches, as in modern swamp cypress, *Taxodium distichum*, which is one of only four living conifers that are deciduous. Two important but now completely extinct species, *Pterophyllum* and *Anomozamites*, formed what we think was a mid-canopy layer. This inference is based on comparisons with modern mid-canopy plant species.

What exactly does "lush, diverse, and complex" mean in relation to vegetation? We considered the word *lush* as both scientists and artists. Artistically, the question of what defines visual lushness was one of the first in a long string that needed to be answered to permit us to attempt an accurate reconstruction. From an artistic perspective, crowding and obvious good health, featuring rich green tones with few yellows and browns, provided a start to depicting lushness. These plants grew across the landscape in a reaching, striving, competitive environment, where, in this time before flowers, the color green would have ruled. In drafting our landscape reconstruction, we took great care to investigate the myriad probable variations in green tones, even comparing the thicknesses of the fossil cuticles with those of living plants so that we could closely predict the green tones that would have prevailed. Thick evergreen leaves, such as those of monkey puzzle and kauri pine, are usually a deeper green than leaves with a short life span, like those of larch, swamp cypress, and dawn redwood, which are bright acidic green. Marlene produced two beautiful and identical watercolor palettes of greens so that we could communicate "green" via email using a handy key for the different hues.

Another artistic aspect of lushness was a pervading sense of abundant water. At first sight, the glistening pendulous droplets hanging from the tips of *Lepidopteris ottonis* appear somewhat incongruous. Why aren't they present on the other foliage? Are they the last relics of early morning dew? In fact, these are guttation droplets, formed at night and composed of sap exuded from the plant itself. Some plants, particularly vines and other climbers, draw large quantities of water up through their roots when soil moisture levels are high. This process creates a positive pressure gradient and forces sap through structures known as hydathodes, or "water pores," at the leaf tips. These microscopic structures were found on the fossil leaf tips of *Lepidopteris*, providing the concrete data we needed to reconstruct this species as a vine rather than a shrub or tree.

The fabric of the sediments also pointed strongly to an abundance of water and frequent flooding events. Although scientists rarely use the word *lush* in a formal biological sense, in the context of Greenland at this precise time

FIGURE 2.4. Marlene Hill Donnelly, *A Palette of Green*. Watercolor on card.

in Earth history, it encapsulates two important biological aspects of the Late Triassic ecosystems: high productivity and high diversity. We estimated that there were at minimum 42 different species of plants present in the oldest plant beds, and that this diversity was stable for many hundreds of thousands of years.

A FOREST CANOPY

Modern forests have distinct zones or "layers" of growth above the ground, which vary over space and time and determine the overall productivity and diversity of the forests as a whole. These include an understory layer, which is the zone of growth closest to ground level and generally dimly lit; a canopy layer, which can range from 10 to over 35 meters above ground level; and finally, an emergent layer of sparser growth. The emergent layer is the loftiest forest stratum. It is an infrequently occupied space, pierced by only the largest and most mature trees in tropical climates, which generally grow to heights of 30–90 meters above the ground surface. Cooler-climate forests like those in New England, New Zealand, and Ireland do not have an emergent layer. Each forest layer has a unique

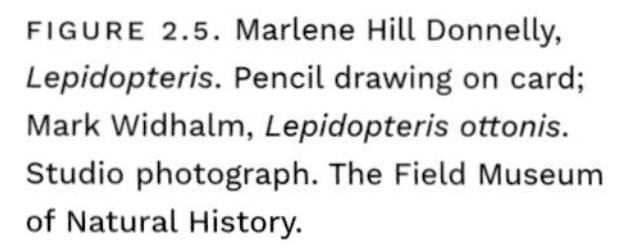

FIGURE 2.5. Marlene Hill Donnelly, *Lepidopteris*. Pencil drawing on card; Mark Widhalm, *Lepidopteris ottonis*. Studio photograph. The Field Museum of Natural History.

FIGURE 2.6. Marlene Hill Donnelly, *Forest study*. Watercolor and ink.

microclimate and light quality, from the dark, humid understory to the sun-drenched emergent layer. Together, these layers create four-dimensional complexity in a landscape over space and time (the fourth dimension!) in which other species can thrive. Indeed, the livelihoods of other plants and animals that occupy the different layers of forests depend on the microclimate variations that the forest trees themselves engineer. In both modern and ancient forests, the number of layers within the forest canopy is an important feature distinguishing between the warm tropics and cooler temperate latitudes. Tropical forests are characterized by intense competition for light because all other environmental conditions for their growth are nearly perfect. Vines and lianas have an optimized strategy to compete for light because they can reach the canopy tops without investing in energetically expensive structures like woody trunks. Competition for light has resulted in the evolution of species that occupy different light environments, or niches, from the forest floor all the way up to the canopy. Tropical forests thus have the globally unique distinction of being highly multilayered. Temperate forests, on the other hand, which today occupy the latitudes north and south of the equator between 30° and 60°, never achieve equivalent multilayered complexity. In general, they also have fewer species of vines, lianas, and epiphytes, which are plants that

live on other plants. This distinction provides the paleobotanist with one means of assessing the "tropicality" of ancient forests. Using fossils to interpret the canopy structure of past landscapes is, however, by no means a simple task.

The most fortuitous fossil plant deposits preserve the whole ecosystem. In these deposits, we find whole fossilized tree trunks upright in life position, which enables confident assessment of the original living tree height. These deposits are called standing fossil forests. We use a mathematical relationship between trunk diameter at chest height and total height of living trees and apply this formula to fossil trees of known girth to estimate the height they attained when part of a living forest. In such situations, it is possible to infer how many canopy layers existed within the ancient canopy as a whole. Multilayered canopies with emergents in addition to a shrub layer and an understory layer indicate a tropical climate.

Standing fossil forests are, however, extremely rare. In the case of East Greenland, we inferred that there were three distinct canopy layers by carefully reconstructing the architecture of each individual fossil species. We used comparisons with modern forest species as an architectural guide. In addition, we compared the Greenland fossils with other fossil plant finds of similar age from different parts of the world. In the Triassic of Greenland, one of the kings of the forest was *Ginkgoites*. From its lofty position of high-canopy dominance, it would have shaded the leaves of the *Lepidopteris* vine, which in turn would have shaded the mid-canopy plants *Anomozamites* and *Pterophyllum*.

From an artistic viewpoint, the canopy layers radiate 360°, representing an amazing compounding of life. Intensive sketching in the rich living forests of the Pacific Northwest and Hawaii indicated to Marlene that far more layers of plants

were required than she had originally assumed to achieve authentic visual depth. The plants' arrangement within the Triassic forest reconstruction required careful attention, as it implied ecological relationships between plants and environment. Some could potentially have their feet in water, like the swamp-cypress look-alike conifer *Stachyotaxus*, while others, like the broad-leaved conifer *Podozamites*, had to be on higher ground, according to clues extracted from the sediments.

FIGURE 2.7. *Facing:* Marlene Hill Donnelly, *A Forest Canopy: A Portal into the Late Triassic.*

FIGURE 2.8. Marlene Hill Donnelly, *Dictyophyllum model.* Paper, adhesive wire, and bonsai wire.

FERNS ABOUND

A striking feature of the Triassic forests is the unusual ferns, with their horseshoe-shaped stems and saw-toothed margins, growing among the litter of the forest floor. These ferns, which belong to the genus *Dictyophyllum*, were one of the most easily identified components of the paleoflora when we unearthed them as fossils from the freezing cliffs of Astartekløft. Every fossil fern leaf prized from the sediments was stamped with a blackened, irregular mesh pattern that revealed where living veins once supplied ample water to a large ferny leaf. Our fossil dictyophyllums very closely resembled the modern tropical fern *Dipteris*, a member of the fern family Dipteridaceae, in both shape and habit. For these reasons, we chose *Dipteris* as our modern analog—our "life model"—for fossil *Dictyophyllum* to guide the artistic rendering of the fossils as living plants. Today, *Dipteris* ferns grow in the tropics from Australia and Asia to the Caribbean, although they are now rare. They are sometimes given the common name umbrella ferns, which, like their Latin name (from *di-*, "two," and *-pteris*, "fern"), is a nod to their umbrella-spoke shape. The young leaves of this genus are particularly striking in their unfurling process: the new leaflets first unroll from a curl, or crozier, then hang down declined and pendulous, flushed with purple and red pigmentation, which can persist until the plant is fully unfurled; such immature fronds may be seen in the foreground below.

UNDERSTORY AND REGENERATION

In our portal view of the dark and still forest understory, we get a glimpse into Greenland's Late Triassic past. Red-tinged shoots of *Dictyophyllum* poke through mature radiating fronds. Nineteen percent of the fossils found in the Triassic Halfway Bed belonged to *Dictyophyllum*. Together with the sedimentary clues in this bed, this suggested that *Dictyophyllum* was an important component of the exposed floodplain following the retreat of floodwaters, but was also present, albeit in lower abundance, in the understory

FIGURE 2.9. Marlene Hill Donnelly, *Understory and Regeneration: A Portal into the Late Triassic.*

of the drier gallery forests that fringed the riverbanks. We think that *Dictyophyllum*, like its modern cousin *Dipteris*, played an important ecological role as one of the first colonizers of disturbed ground, quickly invading spaces left bare by floodwaters.

Marlene had the ingenious idea of incorporating different life stages of the Triassic plants into the reconstruction in order to evoke the feeling of a healthy, regenerating forest, even though some relevant fossils were unavailable. While we had lovely specimens of mature *Dictyophyllum*, we had no fossils of youthful emerging fronds for this species. Instead, our life model *Dipteris* was used to capture the unique unfolding process of these umbrella-shaped fern species. Marlene began by sculpting a model using *Dictyophyllum*'s leaf structure shaped in the known unfolding growth pattern of the living *Dipteris*; she followed with sketches and paintings.

INTERACTION

Insect body fossils from the last 400 million years of Earth history are less rare than one might expect, given their delicate nature. They turn up frequently in fossil tree resin (amber) and in exceptional fossil localities called *lagerstatten*. Given the exceptional quality of fossil plant preservation at our field site, we expected, and indeed hoped, to unearth a few fossil insects, or perhaps even the odd vertebrate fossil. Our searches were, alas, not fruitful. But there are other ways of studying past insect diversity. Just as graffiti on city walls leaves the mark of the street artist, telltale evidence of the presence of insects can be detected in the deep past even in the absence of insect bodies. Insect feeding and egg-laying behavior can leave traces that can be immortalized in fossil leaves. Leaf herbivores, for example, leave behind very diagnostic behavioral "footprints" such as leaf galls, leaf mines, and leaf-margin feeding traces.

In the ancient forests of East Greenland, we found evidence for streamside dragonflies in microscopically sized and exquisitely preserved insect egg-laying (oviposition) scars in *Ginkgoites* leaves and in those of their relatives *Sphenobaiera* and *Baiera*. These scars were more common in the Triassic than the Jurassic beds and were distinctive enough in shape to be assigned to their own fossil group, or "ichnogenus," *Paleoovoidus*. *Ichnofossil* is a term we use for all traces of former life, such as burrows, footprints, limb prints, bite marks, and so forth. The Boundary Bed preserved two ichnospecies of leech cocoons, called *Dictyothylakos pesslerae* and *Pilothylakos pilosus*. Perhaps Greenland's ancient forests were like those of modern Australia, where ground leeches soon sense your warm ankles for a feed if you are not wearing hiking boots. Other expeditions to similarly aged fossil plant localities have discovered fossilized dragonfly pupae sandwiched between the upper and lower leaf surfaces of broad-leaved conifers such as *Podozamites*, which was a vital component of the canopy in the Triassic forests of Astartekløft.

Overall, we think that the forests of East Greenland may have been rather unpalatable to insects compared with many modern ecosystems. Only a handful (less than 1 percent) of over 4,000 fossil plants show evidence of insect feeding damage. Compare that with modern tropical and temperate forests, where we see up to 15 percent of leaves with evidence of insect munching. Insects must have been present in some numbers, however, as the protoflowers of *Anomozamites* and *Pterophyllum* have structures that produced nectar, the sugary substance that attracts the insects necessary for fertilization. Both of these species belong to an extinct order of plants called Bennettitales, which had flowerlike structures, cycad-like leaves, and yucca-like trunks—a rare combination of features not seen in any modern living plant.

FIGURE 2.10. Marlene Hill Donnelly, *Berm study*. Watercolor and ink; *facing:* Marlene Hill Donnelly, *Interaction: A Portal into the Late Triassic*.

There is fossilized evidence for plant-insect interactions dating all the way back to at least the Early Devonian, around 400 million years ago. Insects were clearly well established by the Late Triassic, almost 200 million years later. However, the coevolution of insects and plants, in which one group exerted an evolutionary influence on the other, driving increases in both groups' biodiversity, really exploded after the evolution of the flowering plants in the Cretaceous, around 120 million years ago. It is captivating that some of the first hints of this powerful coevolutionary relationship can be observed in the fossilized proto-flowers that literally littered the cliffs of our field sites in Greenland. In our full landscape reconstruction of the Late Triassic, we acknowledge this early plant-insect relationship with an illustration of a Late Triassic dragonfly, *Italophlebia*, perched on top of the closed proto-flower of *Anomozamites*.

Of course, interesting interactions with plants do not involve animals exclusively: the interactions that vines, lianas, and other climbers have with their plant hosts are equally fascinating. The fossil genus *Lepidopteris* is thought to have been a liana. Modern examples of this life-form include *Clematis*, *Bougainvillea*, and, of course, grapevines. Today, lianas are typically found in deciduous seasonal and wet tropical forests, but occur in other settings, too. Lianas are typically rooted in the soil, have long, woody vine-like stems, and grow using other plants for support to gain access to the canopy, where light is more plentiful than on the dimly lit forest floor. While lianas are not parasites (they rely on their hosts only for support), they can grow very rapidly, as they can devote less energy to woody stems and more to softer and less nitrogen-demanding parts, such as leaves and non-woody stems. This cheeky and clever strategy produces huge problems for their host. The growth of the liana's host is often impeded, and its reproduction is often retarded. Ultimately, lianas strangle their hosts slowly and persistently, which leads to tree deaths and altered forest structure.

It is unsurprising, therefore, that some trees have adaptations that reduce their susceptibility to these stranglers. Smooth or scaly, easily shed bark is one such adaptation. Our living analogs for *Podozamites* included species within the Southern Hemisphere conifer genus *Agathis* (kauri pine). Almost all *Agathis* species have smooth or shedding bark to repel climbers. If *Podozamites* is an ancient relative of living *Agathis*, that shedding bark would have been an excellent defense against the advances of the *Lepidopteris* vine. The mid-canopy plants of the Late Triassic forests have all been illustrated with Hawaiian skirt–like layers of persistent dead leaves hanging down from their stems. In modern ecosystems, these "skirts" are thought to be another possible adaptation to deter climbing plants.

STILL LIFE

Most dioramas and landscape reconstructions of the geological past focus on the enigmatic carnivore of the time, surrounding it with ambiguous vegetation that is difficult to pinpoint to a particular era. Our portal view of the Late Triassic forest floor turns this usual paradigm on its head: the vegetation is rendered in detail, and is of its time and place. At first glance, the landscape appears to be devoid of any predators. There is, however, a small trace of the great dinosaurs that are known to have roamed this part of the world in the Triassic: a theropod dinosaur footprint hides amid the forest floor litter in the foreground.

Reconstructions of the past are also usually shown as pristine, like tidy gardens or still lifes, with every leaf and twig perfectly formed. Death and decay are often not considered, yet both are functions vital for restoration within ecosystems, as both return essential mineral nutrients such as nitrogen, phosphorus, and potassium to the soil. It is difficult to understand the decay of vegetation in the deep past, but not impossible. One very exciting and emerging method is the study of the economic properties of fossil leaves. Modern ecological research has shown that economic theory can be used to categorize

FIGURE 2.11. Marlene Hill Donnelly, *Still Life: A Portal into the Late Triassic.*

every individual plant species on this planet according to the structural and functional characteristics of its leaves. There is a continuous spectrum of leaf economies, ranging from thinly and cheaply constructed leaves, which have high nutritional value and are easily degraded, to those that are thick, tough, energetically expensive, and long-lived. These tough leaves have low nutritional value and do not give back the essential elements of their ecosystems quickly. Simply put, they strongly resist decay. Our studies of the economic properties of the leaves from East Greenland have just started. To date, they suggest a range of leaf types, from easily degradable (such as those of *Anomozamites*) to highly decay resistant (like *Ginkgoites* and *Sphenobaiera*).

The fact that fossil plants have been preserved for approximately 200 million years attests to an environment in which complete decay was prevented for some leaves that fell to the forest floor and into the neighboring river. Following death or shedding, the leaves that went on to be fossilized must have been buried rapidly and removed from an environment where decay would occur. Rapid burial would also have limited the physical shredding and breakdown that can happen as leaves are transported by wind or rivers and are altered by burrowing and rooting organisms. Vast quantities of sediment dumped during flooding events provided the means for this rapid burial and enabled the preservation of the Astartekløft fossils that has given us such a clear window into Greenland's tropical past.

Marlene used our analyses of leaf economies, combined with modern plant analogs, to bring a decaying Triassic forest floor to full color. Fallen *Ginkgo* leaves and the twisting spent leaves of cycads in her studio morphed subtly into their far-distant ancestors and drifted back in time. Many layers of fallen leaves needed to stack up in Marlene's sketches and final full-color illustrations to convey both visual depth and the feeling of a soft, moist forest floor through which great predatory dinosaurs trod in lethal silence. Her field studies of modern floodplain environments became an important source of inspiration.

FIGURE 2.12. Marlene Hill Donnelly, *Study of decay*. Pen and ink on card.

LONGEVITY

Just as individuals have a finite life span, so, too, do species and families. On average, species survive for about 2 million years before going extinct. One of the most remarkable aspects of plant families, compared with animal families, is that so many of them have exceptional longevity. The poster child of the plant world for longevity is *Ginkgo biloba*, the maidenhair tree, which you may know as "stinko ginkgo," a common urban street tree with a unique odor. As an individual, it is thought to live for up to 4,000 years. As a family, it has a tremendously deep fossil record stretching back to the Triassic. *Ginkgo biloba*'s Late Triassic relative was *Ginkgoites minuta*. We featured it prominently in the canopy in our reconstruction, as its leaf fossils were abundant in the Triassic fossil beds. The fossil leaves, like their modern relatives, bear the trademark bilobed leaf form: the leaf blade is dissected into two, more or less even, fan-shaped parts. *Ginkgoites* flourished in forests across the entire East Greenland landscape in the Late Triassic, but seemingly went locally extinct as global temperatures and carbon dioxide concentrations in the atmosphere rose across the Triassic-Jurassic boundary.

Although global climate warming at the Triassic-Jurassic boundary did not cause global extinction of all ginkgos, it resulted in the loss of most *Ginkgoites* trees from the forests of Jameson Land and neighboring Sweden for many hundreds of thousands to millions of years. We believe that some small populations of *Ginkgoites* trees survived the climate and environmental changes of the Late Triassic and earliest Jurassic. Most perished locally, however, and the group survived only by migrating toward the North Pole, where the prevailing climate was cooler. Some of the earliest signs of the biological impacts of modern climate change have been shifts in the biogeographic ranges of hundreds of plant and animal species toward the cooler poles and cooler reaches higher up mountains as their local climate zones have heated up. Just as climate change is forcing the migration of people around the world, it is driving the migration of plant species that do not have the capacity to adapt their temperature tolerances. Plants, of course, cannot uproot themselves and walk, swim, or fly to a new home range. Instead, each species casts its seeds and spores into the wind, or onto the body of an insect or vertebrate herbivore, and migrates slowly with each new generation. If the pace of climate change exceeds the rate at which this slow intergenerational migration can occur, then the species as a whole is at high risk of going extinct.

In the case of Triassic *Ginkgoites*, we believe that the pace of global warming did not exceed the plants' capacity to migrate northward, because they

FIGURE 2.13. Marlene Hill Donnelly, *Longevity: A Portal into the Late Triassic.*

FIGURE 2.14. John Weinstein, *Ginkgoites minuta.* Studio photograph. The Field Museum of Natural History; Marlene Hill Donnelly, *Ginkgoites minuta.* Watercolor on card; *facing:* Marlene Hill Donnelly, *Sketch of modern ginkgo tree.* Watercolor and ink.

survived. Those that survived adopted a different leaf economy, as their leaves became thicker, tougher, and more stress tolerant. We predict that many contemporary plant species will try a similar strategy as weather patterns and climates shift due to human-driven global warming. We are now facing both climate and biodiversity emergencies that are unfolding at faster rates than those that occurred across the Triassic-Jurassic boundary. Later in the book, we will examine whether, and how, species and ecosystems adapted to Late Triassic climate change, and what lessons they can offer for dealing with our contemporary crises. For now, we need to establish the baseline: What did the Triassic ecosystems look like, and how did they function, before global climates warmed dramatically? Perhaps more importantly, how did some plant families, like the ginkgos, manage to survive and persist in the face of such extreme climate change?

Marlene was equally fascinated with our *Ginkgoites* fossils. She captured their almost otherworldly longevity by observing modern and fossil ginkgos side by side in her studio. She viewed the colorless cells of their leaf margins as tiny lenses, catching the light to form a miniature halo for every leaf. These cells are actually the stem cells of ginkgo leaves, located exclusively on the leaf's outer margin. They have the capacity, unique among leaf cells, to produce new cells as the leaf grows.

The cycad family—comprising plants with stout woody trunks and ever-

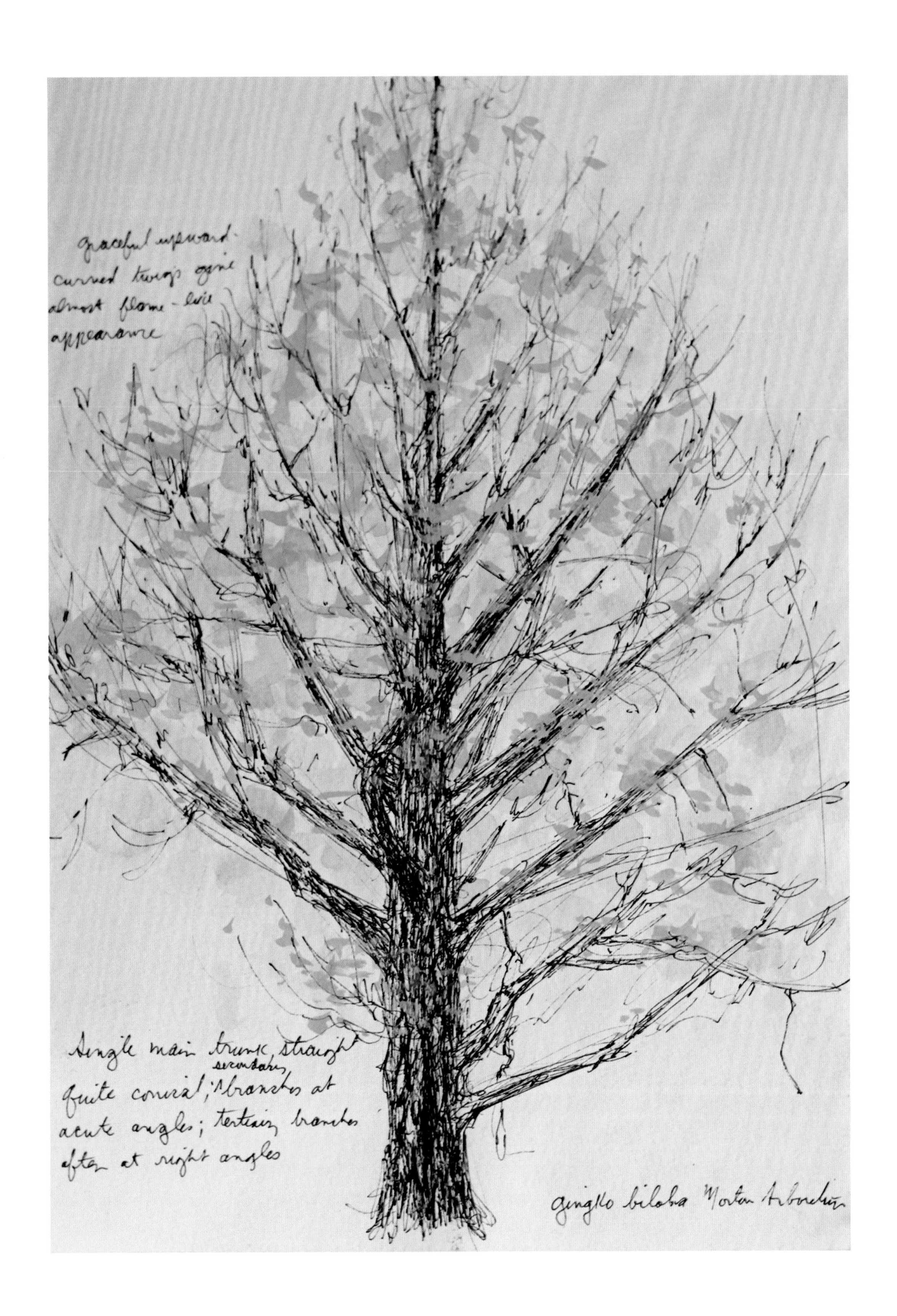
graceful upward-
curved twigs give
almost flame-like
appearance
Single main trunk, straight
secondary
quite conical; branches at
acute angles; tertiary branches
often at right angles
gingko biloba Morton Arboretum

FIGURE 2.15. John Weinstein, *Pseudoctenis*. Studio photograph. The Field Museum of Natural History.

green palmlike leaves—also has a very long fossil record. The Jurassic is often called the Age of Cycads. However, this lineage can be traced even further back, to at least 300 million years ago, during the late Carboniferous, and its fossils have been found on every continent. Today, cycads are generally restricted to the latitudes between 30° N and 35° S (think of the latitude for Cairo, Egypt, or Shanghai, China, in the north and Buenos Aires, Argentina, or Sydney, Australia, in the south), but are most diverse in the subtropics. These plants are found in a wide range of environments, from semidesert to wet rainforest, and can even grow in oxygen-poor bog-like soils. Some species are salt tolerant, and some can tolerate freezing weather and moderate annual snowfall. During the Triassic in Greenland, the cycads included *Doratophyllum*, *Ctenis*, *Pseudoctenis*, and *Nilssonia*. Their longevity as species at the Astartekløft locality was limited, as they are not found in the Jurassic, where they appear to have gone locally extinct. In contrast with the ginkgos, cycads were not resilient to the dramatic climate changes that unfolded during this interval. They, too, disappeared from Greenland, but did not come back and recolonize when global climates cooled. The reasons remain unknown to this day: perhaps they could not migrate at the same rate as *Ginkgoites*, or were unable to establish themselves in new home ranges because of competition from other plant species. The relative rarity of fossil cycads at Astartekløft made it particularly thrilling to unearth them. As we excavated our way through the Astartekløft sediments, we began to find fewer and fewer of their huge dissected leaves with their big square leaflets, and concluded that the species they belonged to were going extinct as global temperatures heated up.

COMPLEXITY

The word *complex* derives from the Latin roots *com-* (together) and *-plex* (plaited, woven). There is no scientifically agreed-upon definition of *complexity*, but in ecological studies, the term typically refers to the many interactions among living organisms and their environment: more complex ecosystems exhibit greater numbers of these linkages. Ecological complexity can range in scale from the molecular to the planetary, and the connections may be behavioral, biological, chemical, environmental, and even social or cultural. More simply put, this "biocomplexity" is the complete web of interactions that connect all parts of an ecosystem.

We attempted to quantify the ecological complexity at play in the Triassic forests by examining the behavior of key forest species. We focused specifically on plant behavioral and physiological characteristics that could exert an influence on other parts of the ecosystem. For example, counts of stomata and veins on the fossil leaves of *Ginkgoites* and *Anomozamites* revealed that the latter probably had much higher rates of gas exchange, and therefore higher photosynthetic rates, than *Ginkgoites*. All plants use their stomata to take up carbon dioxide gas from the atmosphere, which they use as a carbon source to manufacture sugar during photosynthesis. When stomata are open to take in carbon dioxide, however, water vapor is lost at the same time. Plants must constantly optimize this exchange of gases—carbon dioxide in and water vapor out—so that it is finely tuned to the benefit of the

FIGURE 2.16. Marlene Hill Donnelly, *Complexity: A Portal into the Late Triassic*.

plant. One strategy for optimizing gas exchange is controlling the total number and size of stomata on the leaf surface. Our studies revealed that *Anomozamites* was able to gain more carbon per unit time than *Ginkgoites* because it had far more microscopic stomata on its feathery leaves, but that this gain came at the cost of losing much more water. This demonstrated to us that a complex mosaic of species with different leaf shapes, architectural forms, *and* physiologies made up these ancient forests: some were water wise (like *Ginkgoites*), and others were big water spenders (like *Anomozamites*), just as in modern forests and woodlands. Ongoing research and experiments aim to predict which living species and ecological strategies will be the winners and losers as global climates warm. Will they be water spenders with high photosynthetic rates, or species that use resources such as water more conservatively? One of the drivers of our own study was the desire to address this question from the perspective of the Astartekløft fossil forests. As global climates warmed toward the end of the Triassic period, would the fast-paced *Anomozamites* or the more conservative *Ginkgoites* be the ecological winner in the hotter world of the earliest Jurassic?

FERTILITY

We think that vegetated Triassic landscapes were predominantly green, with few injections of reds, oranges, blues, yellows, and violets. The terrestrial flora of the time was almost exclusively populated by plants with decidedly green or brown reproductive parts. Most plants of the Triassic used wind for their reproduction, so color was not needed to attract other organisms for fertilization. Pigments in plants, or the nanoscale structure of plant surfaces, alter the way the sun's light is reflected and absorbed, resulting in the colors we perceive. Those that result in reds, purples, and other non-green hues have evolved mainly as defenses against sunburn, which can cause genetic mutations, or as attractants for animals that are deployed by the plants to cross-fertilize their male and female reproductive structures. Conifers, ginkgos, cycads, and other common Triassic plant groups produced their seeds in relatively demure structures, such as cones. Flowering plants—those show-offs that brought such an explosion of color to the world in both their flowers and fruits—had not yet evolved. We had to imagine a world devoid of flowers and fruits in all their beautiful hues. Our artistic reconstruction of the Late Triassic forests is therefore a celebration of green, with only hints of orange, brown, and rust to illustrate the youthful or senescent stages of leaves, and blacks and silvers to illustrate the vine-repelling bark of trees.

Although there is a scientific method of determining color from fossil pig-

FIGURE 2.17. Marlene Hill Donnelly, *Fertility: A Portal into the Late Triassic.*

ment cells called melanosomes, it has not yet been widely applied to fossil plants. We therefore took a conservative approach to color, even in the case of the rendering of the closed *Anomozamites* proto-flower above, illustrating it in a rich brown tone to depict the toughness of its protective bracts, which we found in such great abundance in the Flower Bed at Astartekløft.

Why is it that, amid today's green plants, only the true flowering plants (angiosperms) are highly colored? Why are conifer cones mostly brown and only sometimes purple or a dull red? It all probably comes down to *coevolution* again: a term first coined in 1964, meaning reciprocal evolution between interacting species. The classic example of coevolution is that of insect pollinators and flowering plants, remarked on as far back as 1859 by Charles Darwin in his *On the Origin of Species*. Flowers and flower parts are colored in both the visible and ultraviolet (UV) portions of the spectrum to attract insects for pollination and seed dispersal. Humans perceive light energy of wavelengths from 400 to 700 nanometers—the visible spectrum—and have three kinds of photoreceptors in the eye, red, blue, and green (RBG); that is, we are trichromatic. Bees are trichromatic as well, but they perceive light energy of wavelengths from 300 to 600 nanometers and have blue, ultraviolet, and green (BUG) photoreceptors, which allow them to see in colors ranging from yellow-orange (but not red) to ultraviolet. Flowers reflect light across the spectrum; we perceive the colors red through purple, while bees perceive yellow through ultraviolet. This means that a flower's appearance is very much determined by the eye of the beholder. However, plants also fluoresce, meaning that they absorb ultraviolet light, then re-emit it at longer wavelengths visible to the human eye. This fluorescence causes us to perceive the flower as lustrous and glowing. The ultraviolet display of flowers to attract their pollinators has been honed to spectacular effect by tens of thousands of flowering plant species. It is likely, however, that UV fluorescence first evolved in the spores and pollen grains of Earth's most ancient plants to protect the sex cells within from DNA-damaging UV radiation.

The dragonfly in our reconstruction of the Late Triassic is perched on top of an *Anomozamites* proto-flower. There is no consensus on the pollination biology of this extinct group, and we do not currently have sufficient scientific data to reconstruct its true color. However, the possible presence of nectaries and insect-trapping devices in some *Anomozamites* and their relatives within the Bennettitales indicates that they may have needed insects for their

FIGURE 2.18. Marlene Hill Donnelly, *Color and Light: A Portal into the Late Triassic.*

pollination. As for the pollinating insects themselves, beetles in the family Cupedidae are among the leading candidates.

At Astartekløft, we found a mass occurrence of a fossil called *Cycadolepis* in the Flower Bed. These fossils are the protective wrinkled bracts of proto-flowers that belonged to *Anomozamites*. The tremendous number of fossil flower parts in this bed suggests that there was a mass flowering of *Anomozamites* plants at precisely the time floodwaters inundated the plants and covered them with sediment. If these proto-flowers were indeed colorful, it must have been amazing to look across a vast landscape of huge, showy, and most likely long-lasting flowerlike structures, blooming in unison before the catastrophic flooding event that would preserve them in sediments for the next 200 million years.

COLOR AND LIGHT

Marlene's artistic view of the world and Greenland's past kept the scientists thinking about color. What information could we glean from the fossils or, more broadly, through our reading of the literature that would bring the Astartekløft fossils to life and ground them firmly in an unseen time and place? In Marlene's reconstruction of the Late Triassic, rays of sunlight are shown penetrating the canopy and illuminating the multi-hued vegetation below. However, even this seemingly mundane event is not the same today as it probably was in the deep geological past. The sun's rays have become more intense over geological time because all stars, including the sun, age and become more luminous with time (unlike the human aging process—sigh). Prior to around 140 million years ago, solar luminosity was significantly less than it is today, and during the Late Triassic and Early Jurassic, it was about 2 percent less than it is now. But while the Triassic sun's rays would have been less intense, they would have reached Greenland throughout the year. Today, Astartekløft sits at a latitude of roughly 71° N, well within the Arctic Circle, and so is subject to the darkness of a polar winter. In the latest Triassic, though, this region is calculated to have laid about 20° farther south, at a latitude near 50° N, roughly equivalent to that of modern Dublin, Ireland.

There is an integral relationship between the sun's light, leaf pigmentation, and plant physiology. The green pigments in chlorophylls absorb light energy for photosynthesis. Yellow pigments, such as yellow carotenoids (visible in the color of a ripening tomato), contribute to the transfer of energy to the photosynthetic system, but also help to prevent sunburn when the plant absorbs too much light energy, particularly UV radiation. Carotenoids are antioxidants: they mop up free radicals generated by UV light, which are unstable molecules that can damage a plant's genetic machinery. They act like melanin in our own skin cells, which also scavenges free radicals and renders them harmless. Red pigments (anthocyanins and red carotenoids) are thought to have a range of antioxidant and antifungal protective roles. We did not have direct evidence for the color or pigment content of Triassic leaves, but we wanted to reconstruct them as faithfully as possible. To accomplish this, we compared our fossils with appropriate living species, such as the red carotenoid–rich kauri pine (*Agathis australis*), our chosen analog for *Podozamites*, to infer their likely hues. It is an imperfect method, but the best we could do with the available fossil data. A future step will be to chemically characterize fossilized leaf pigment molecules to infer their color, which is an exciting avenue of unexplored research.

ADAPTATION

To reconstruct these lost landscapes, we relied in equal parts on the fossil plant discoveries and on detailed study and interpretation of the sediments from which the fossils were extracted. On the basis of the association of certain species within the Greenland flora with sediments that indicated floodplain conditions, we deduced that a number of taxa were adapted to life in standing water. This characteristic is by no means universal among plants today. On the contrary, only a relatively low proportion of the world's 350,000 or so plant species can withstand the low soil oxygen levels caused by flooding. Our interpretation of flood adaptation is also based on comparisons of fossil plants with modern species.

The three taxa illustrated in our portal view of the full Triassic landscape, *Stachyotaxus septentrionalis*, *Anomozamites*, and *Pterophyllum*, appeared repeatedly in floodplain sediments. These sediments were deposited as a result of massive flooding events in the river systems feeding into the great Kap Stewart Lake. Flooding events were common in this area of Greenland throughout the Late Triassic. In addition, the overall structure and shape of *Stachyotaxus* fossil leaves evoked those of the modern swamp cypress (*Taxodium distichum*), a swamp-living giant. Another clue to the soggy habitat was the fact that we found mass occurrences of fossil *Stachyotaxus* seeds. Such mass release of seeds is a common strategy of flood-adapted plants. Although our Triassic reconstruction depicts a nice quiet day sometime after peak floodwaters had receded, the incredible depth of the floodwater deposits—up to 2 meters (which we inferred from the depth of the sediment deposited in each fossil plant bed)—suggests that these events were often violent and may well have been driven by monsoonal rainfall in the Late Triassic.

FIGURE 2.19. Marlene Hill Donnelly, *Field sketches of modern floodplain.* Watercolor and ink on card.

Marlene was unsure how to decode the story of repeated flooding events told by the sediments of Astartekløft. She had no immediate visual frame of reference for a periodically flooding forest, as compared with a swamp where trees live in permanent standing water. Luckily, however, she has the good fortune to live near a floodplain, so following the next heavy rains, she waded into the newly flooded forest and perched on a camp stool for several hours, sketching and observing. Her sketches show that the tree species of the floodplain differ from those just a few yards away on higher ground, and that there are no shrubs or understory—flood-tolerant plants are indeed limited. Instead, untidy tangles of dead logs and branches, swept in by previous floodwaters, claim the ground. Marlene's keen observations of dark rings around tree trunks just above the water were incorporated into the Triassic landscape. They hint at the rapidly dropping flood levels of a raging Greenland river from a time and place that has remained unseen for 200 million years. Now, we want the viewer to get lost in this ancient landscape in the quiet after the floods have receded, and to imagine the soft popping symphony of countless tiny air bubbles, rising out of freshly deposited muds and fine silts as they burst at the water's surface. This is the calm before the storm when volcanoes, climate change, and extinction stamped their imprint into the sediment archives of Astartekløft and the fossil plants entombed within.

FIGURE 2.20. Marlene Hill Donnelly, *Tropical Arctic: A Portal into the Lost Triassic Landscape of East Greenland*.

Marlene Hill Donnelly

3 Crisis and Collapse

Everything changed in the Jurassic—the climate, the atmospheric composition, the vegetation structure and function, the river architecture, and even the fundamental biogeochemical cycles of elements and molecules such as carbon, nitrogen, and water. The fire ecology of the ecosystem shifted, as did the ecological roles of the different plant players. The high-diversity forest and communities of the floodplains collapsed. The chemistry and sediments of the Astartekløft cliffs, and the changing abundances of the fossil pollen, spores, leaves, and reproductive structures that they contained, all pointed at an environmental and biological catastrophe during the transition from the Triassic to the Jurassic period. In fact, the chemistry of the entire biosphere-atmosphere system shifted! We know this because indirect measurements of the carbon content of the atmosphere of this time show a greater than usual abundance of a light carbon isotope, called carbon-12, with six neutrons in its atoms. This distinct chemical signature signified a massive release of greenhouse gases from volcanic sources or from biologically derived methane, both of which are rich in carbon-12. The higher-than-typical concentration of carbon-12 in every carbon-containing compound of the latest Triassic and earliest Jurassic system, including the limestones, the plant leaves, atmospheric carbon dioxide molecules, fossil soils, and fossil shells and teeth, demonstrated unequivocally that a major environmental event had occurred that was triggered by greenhouse gases such as carbon dioxide and methane. To top it all, levels of poisonous mercury, which is also released from volcanic sources, also soared.

Decades of research by multiple teams from different disciplines have unpicked, piece by piece, the nature of the environmental changes across the boundary and their causes. The consensus is that the environmental upheaval was triggered by an intense volcanic episode of exceptionally long duration in the Central Atlantic region—an event with an intensity greater than that of all of modern Earth's active volcanoes combined, from Stromboli to Mount Saint Helens to Kilauea. It progressed in roars and blasts for over 600,000 years and released vast quantities of greenhouse gases, with their telltale carbon-12, into the atmosphere. Today, due to fossil fuel burning and land use change, we are experiencing firsthand what happens when a vast amount of carbon is released into the atmosphere. The global climate warms, oceans become more acidic, and ice sheets and glaciers melt. Weather patterns become more extreme. Species migrate and change the timing of their annual rhythms. Considerably more greenhouse gases were released at the close of the Triassic period into an already ice-free world, so that world experienced a greenhouse warming event of a far greater magnitude than today's. It happened on the order of 10–100 times more slowly than the current climate warming, however. In this chapter, our portal views into the earliest Jurassic world have been rendered to life using fossil species that survived this great environmental crisis.

FIGURE 3.1. Marlene Hill Donnelly, *Volcanic eruption study*. Abstract charcoal on card.

FIGURE 3.2. Marlene Hill Donnelly, *Lava sketch*, Hawai'i Volcanoes National Park. Pigment paint on card.

Standing at the vantage point of Astartekløft camp and looking south toward the cliff section of fossil-rich Triassic and Jurassic rocks, you would see no obvious demarcation of a cataclysmic change. In contrast with other mass extinction boundaries in Earth history, such as the one marking the demise of all non-avian dinosaurs, where in many areas a distinctive sequence of white ash followed by black coal provides a visual full stop for the Cretaceous and a starting point for the Paleogene, no simple and universal sediment markers are available to designate the Triassic-Jurassic boundary. Instead, a combination of stratigraphic tools, all of which require detailed lab-based investigation, is required, including pollen and spore analysis and geochemical study. Application of these tools to Astartekløft revealed that the boundary of the Triassic and Jurassic occurs within the Boundary Bed (Bed 5) at approximately 47 meters up the cliff face above our camp. Geochemical analysis revealed strong excursions in the proportion of carbon-12 at this time, implying a great global flushing of the atmosphere with volcanic carbon dioxide, methane, and sulfur dioxide gases.

Having reconstructed the Late Triassic forests, both scientifically and artistically, in all their lush green glory, we wanted to repeat the exercise for the Jurassic. What did the forests of East Greenland look like during the height

FIGURE 3.3. Luke Mander and Claire Belcher, resin-embedded polished sediments from Triassic-Jurassic boundary, Astartekløft.

of environmental disturbance, when volcanism in the Central Atlantic was raging? Were there indeed forests at this latitude, or did the vegetation structure and makeup change? How could these changes be best represented visually to communicate ecological upheaval and turnover?

DEATH AND SURVIVAL

The transition from the Triassic to the Jurassic belongs to a macabre paleontological group known as the "big five" mass extinctions—and marks the third greatest animal extinction event in Earth history. This boundary was a time of death for individual organisms, for populations, and ultimately for whole species and even families. Around 22 percent of all animal families went extinct. These animal extinctions had multiple causes, such as a decrease of oxygen in the oceans, poisons such as sulfur dioxide and hydrogen sulfide on land, and greenhouse gas–induced global warming, all probably caused either directly or indirectly by volcanic activity in the Central Atlantic region, which at that time was not yet the Atlantic Ocean we are familiar with today.

As the supercontinent Pangea began to break up in the Late Triassic, North America began to separate from Europe and South America from Africa, and the Atlantic Ocean began to form, a massive volcanic episode poured lava onto the land surface and vented volcanic gases into the atmosphere. Over time, the lava would build the flood basalts of the Central Atlantic Magmatic Province—CAMP for short. Atmospheric concentrations of carbon dioxide and methane increased dramatically during this volcanic episode. Carbon dioxide concentrations of about 600 parts per million (ppm) before the boundary climbed to nearly 2,500 ppm in the earliest Jurassic. (For comparison, atmospheric carbon dioxide concentrations today are over 400 ppm and will double by the end of this century if we continue on a "business as usual" economic trajectory.) We calculated that this flux of greenhouse gases raised Earth's average global annual temperature by as much as 7°C (12.6°F) above modern levels, which represents a massive global warming.

Volcanism, and the environmental catastrophe that ensued, extinguished marine reptiles, a soft-bodied eel-like group called conodonts, and sphincto-

FIGURE 3.4. Marlene Hill Donnelly, *Volcano sketches*. Watercolor and ink.

zoan sponges, and caused huge declines in other marine species, including ammonoids, corals, bivalves, gastropods, brachiopods, and foraminifers. On land, many terrestrial vertebrate species were lost to extinction, but crocodiles, turtles, mammals, dinosaurs, and pterosaurs were not as heavily affected. As with all catastrophes, there are always a handful of species that are preadapted to the "new normal"—the disturbed environments that emerge during and in the aftermath of environmental disaster. It is hypothesized that the Triassic-Jurassic boundary crisis created ecological niches for dinosaurs to exploit and enabled their rise to dominance later in the Mesozoic (the geological era that included the Triassic, Jurassic, and Cretaceous periods).

While global plant biodiversity did not change dramatically, plants did not escape unscathed. Our fossils indicated that a great ecological turnover had taken place right where we were standing over 200 million years previously. That turnover was astonishingly clear in the Boundary Bed of Astartekløft, the bed that marked the transition between the Triassic and Jurassic periods, where we painstakingly excavated fossils centimeter by centimeter. In the field, although we moved through time from the oldest to the youngest fossil beds, it is worth noting here that once we started excavating a bed, it was not possible to start in its oldest layers at the bottom. The only way that fossils could be excavated was by starting at the top of the bed, in the youngest sediments, and working backward in time and through sediment layers

FIGURE 3.5. Marlene Hill Donnelly, *Initial sketch of the Disaster Bed*. Charcoal on paper.

to the oldest! This was like playing back time or viewing a film in reverse and trying to make sense of the drama. In the Boundary Bed, a pattern emerged: the fossil species that were most commonly encountered in the uppermost strata of the bed became rarer and rarer as we moved downward. Then, suddenly, in the middle of the bed, different species appeared for the first time and became more and more common as we hammered and chiseled our way down. The almost complete switch in the commonness and rarity of different fossil species—in which the common species became rare and the rare became common—represents the ecological turnover event immortalized in the Greenland cliffs. If this pattern had been discovered in only a single bed at a single locality, it would be inconsequential. However, because the same ecological turnover has been documented in southern Sweden, North America, and Germany as well as in Greenland, this turnover event takes on global significance.

Our excavation of the Boundary Bed confirmed a stark pattern first documented in these cliffs by Tom Harris in the 1930s: over 85 percent of the plant species that were present in the Triassic fossil plant beds of Astartekløft did not make it through to the Jurassic. They went extinct. Many new species emerged, however, meaning that new species evolved. Only one plant family went extinct globally: the Peltaspermaceae, to which our beautiful *Lepidopteris* vine belongs. Thus, for plants, this event, although profound, was not as severe as the mass extinction recorded for animals. The patterns we observed indicated a wholesale ecological reordering. The first sketched view of the Disaster Bed, with its dead trees and its striking absence of live forest, dramatically conveys this ecological turnover. The dominant species of this Greenland landscape, as recorded by their fossils, were ferns and the jointed bamboo-like horsetails. This flora resembled the opportunistic pioneer communities occupying freshly cooled lava flows today on Mount Etna in Sicily. There is always life, even during the most extreme of environmental conditions, but it can take thousands to hundreds of thousands of years for the look of a place, its ecology and species fabric, to recover the ecological order that is lost.

SICKNESS

Our aim with the Jurassic reconstruction was to depict the balance of life and death in the new landscapes that emerged after the Triassic-Jurassic catastrophe. We conducted experiments by exposing living plants to high concentrations of sulfur dioxide similar to those hypothesized to have existed during the CAMP volcanic episode. Marlene sketched the effects of these sul-

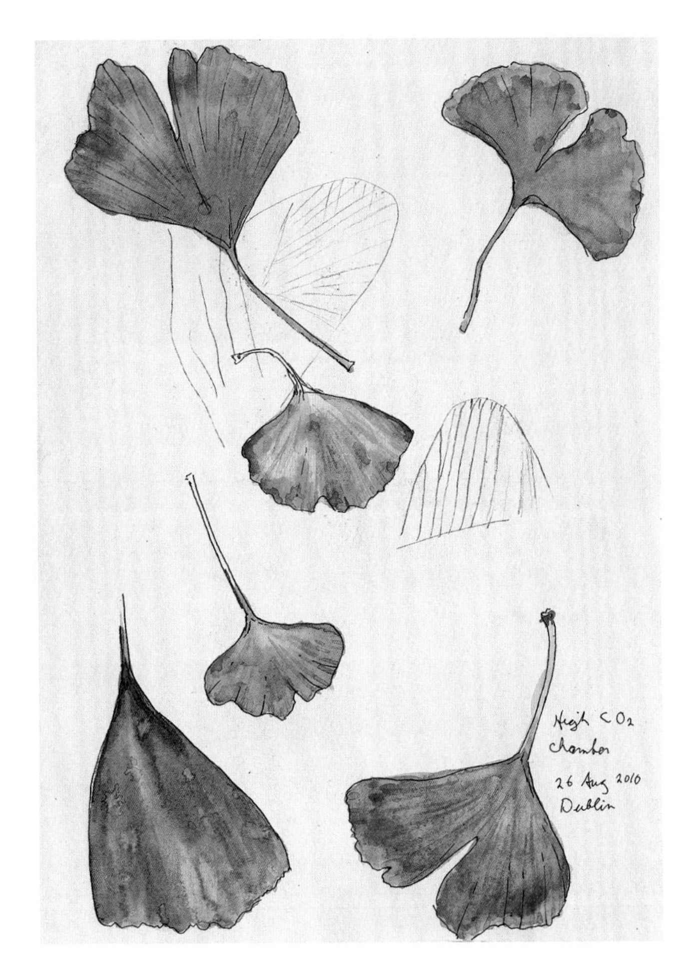

FIGURE 3.6. Marlene Hill Donnelly, *Study of leaf damage in Ginkgo biloba following exposure to elevated sulfur dioxide*. Watercolor.

fur poisoning experiments and incorporated this scientific detail into the reconstruction as distinct patterns of yellowed (chlorosis) and browned dead patches (necrosis) at the leaf tips. Some of our experimental species, like the tough-leaved conifer *Araucaria bidwillii*, were almost impervious to the toxic gas. *Ginkgo* kept repeatedly trying to make miniature leaves, but they shriveled and died before fully expanding and were almost comically small. All of the experimental ferns died within five days of exposure to sulfur dioxide. Our observations of leaf damage and coloration changes in these modern experiments, together with fossil evidence from the Disaster Bed, helped us to peer into the earliest Jurassic world. The Disaster Bed was our only marshy environment in the whole cliff section, and a marsh that was distinctly suffering from a sulfurous acid rain.

The *Elatocladus* conifer foliage in the foreground of the Disaster Bed landscape reconstruction is yellowing and unhealthy. Many over-loved houseplants have this rather forlorn look, as they suffer the effects of chlorosis caused by poor drainage and overwatering. Chlorosis results when a plant produces insufficient chlorophyll. This in turn leads to inefficient photosynthesis and an inability to manufacture carbohydrates. Plants that are adapted to swamp and marsh conditions rarely show chlorosis, however, as they have developed mechanisms of ensuring that their roots maintain good oxygenation under even the soggiest and most anoxic conditions. We made the decision to depict chlorosis in the *Elatocladus* foliage of our portal view as a visual reference to data from many laboratories around the world, whose research implicates high concentrations of sulfur dioxide in the Triassic-Jurassic mass extinction event. The geologists, paleontologists, and geochemists who have studied this extinction boundary in detail generally agree that the extinction was caused by a great confluence of environmental changes, all triggered by CAMP volcanism.

Sulfur dioxide became infamous in the 1970s when emissions of the gas from power plants were implicated in acid rain. The primary pathway by which atmospheric sulfur dioxide enters plants is through their stomata, though to a much lesser degree, it can also penetrate the plant cuticle. When plants are exposed to the gas at exceptionally high levels, its negative effects on photosynthesis and energy metabolism are chronic, resulting in injury through reduced growth and yield as well as through premature leaf drop.

The Greenland fossil flora preserves evidence for the effects of the volcanic outpouring of sulfur in the shapes and sizes of its leaves. Paleobotanist Karen Bacon's experimental studies on living plants in simulated Triassic-Jurassic boundary atmospheres show that leaf shape becomes rounder following exposure to just 2,000 ppb of sulfur dioxide. This concentration is lower than was typical of many industrially polluted cities in the 1980s, but high enough that modern legal systems in most countries now allow only around 4 hours per day of human exposure to similar levels. Fossil leaves from the Flower Bed at Astartekløft, just prior to the Triassic-Jurassic boundary, show a pronounced increase in leaf roundness, suggesting that the toxic gases had already started to be released from CAMP volcanoes at this time.

RESILIENCE

Nature is remarkably resilient to change in environmental and climate conditions if given sufficient time. We see this manifested in the East Greenland fossil flora. Amid the environment-destroying force of the CAMP volcanic episode, which altered the composition of the atmosphere, caused ocean acidification by increasing the amount of carbon dioxide dissolved in seawater,

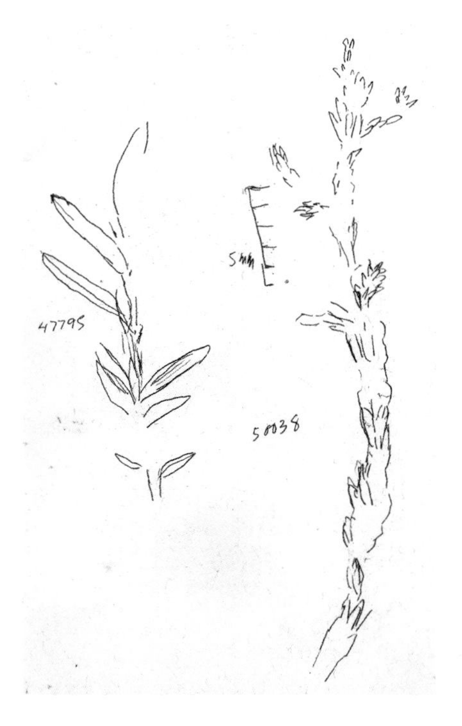

FIGURE 3.7. Marlene Hill Donnelly, *Sickness: A Portal into the Earliest Jurassic*; Marlene Hill Donnelly, *Elatocladus*. Pencil drawing.

decreased the ocean's oxygen content, and changed the entire energy balance of Earth through the action of heat-trapping greenhouse gases, *some* species still survived. How was this possible? The environmental changes across the Triassic-Jurassic boundary occurred at a sufficiently moderate pace to enable some plant species to adapt to the new environmental conditions. Additionally, some of the Jurassic survivors were unusual in having an exceptionally wide range of ecological and environmental tolerances. They were essentially the weeds of their time. They were able to eke out an existence almost anywhere and to rapidly take a foothold where other species were losing ground.

Notable among these resilient survivors were the horsetails—represented by our portal view of a small stand of *Equisetites*. They look remarkably similar to modern horsetails, and at one point geologists would have placed them in the living genus *Equisetum*. These easily identified plants are clearly also extremely resilient—they have persisted in the fossil record for over 300 million years and have remained largely unchanged in overall body plan and structure for at least the last 200 million! We know from modern studies of *Equisetum* that the plants are well adapted to survive in a range of habitats, from woodlands and fields to freshwater shorelines. The living genus also has an almost global biogeographic range, as it occurs from the Arctic to the tropics and from the Himalayas to the Andes, and is absent only from Antarctica. It is therefore unsurprising that *Equisetites* fossils fell out of the rocks as we cracked them with hammers in our coaly excavation pits in the Disaster Bed.

FIGURE 3.8. Marlene Hill Donnelly, *Resilience: A Portal into the Earliest Jurassic.*

OPPORTUNITY

Climate change and atmospheric pollution removed the trees that had dominated the latest Triassic landscape for millions of years. Ferns were able to exploit this opportunity and explode into this newly vacant niche. This changed pattern of dominance was one of the most remarkable features of the Disaster Bed compared with the older fossiliferous layers. The fact that a proliferation of ferns, called a "fern spike," was also observed in localities in Germany, North America, and elsewhere indicated that this feature was very much a global phenomenon.

Standing in a modern savanna, prairie, pampa, or steppe, an observer would see grass as the predominant vegetation. In the earliest Jurassic of Greenland, that observer would have seen ferns almost wherever the eye looked. The global fern spike is better documented by palynology, the study of pollen and spores preserved in sedimentary rocks, than by the fossil leaf record. Fern spikes are often associated with perturbations of the environment, with perhaps the most famous fern spike of all following the mass extinction at the end of the Cretaceous, caused by a meteorite impact.

FIGURE 3.9. Marlene Hill Donnelly, *Opportunity: A Portal into the Earliest Jurassic*; John Weinstein, *Thaumatopteris brauniana fossil from Boundary Bed*. Studio photograph. The Field Museum of Natural History; Marlene Hill Donnelly, *Thaumatopteris brauniana model*. Copper and bonsai wire.

Plants reproduce by means of spores and pollen grains. These tiny microscopic bodies, sometimes as small as 5 micrometers (using imperial measurements, that's about two ten-thousandths of an inch) in diameter, are often widely transported. Due to their incredibly resistant coating of a polymer called sporopollenin, they can be preserved in sediments laid down by rivers, lakes, deltas, estuaries, bogs, and even oceans for hundreds of millions of years. Because of their high preservation potential and wide distribution, palynologists can study and count pollen and spores to understand ancient vegetation. In Greenland, we processed 40 samples across all our fossiliferous beds for pollen and spores, and in each, at least 350 pollen and spore types were observed, counted, and meticulously recorded by then PhD student Luke Mander. Hours of microscopy later confirmed the fern spike suggested by the fossil leaf data. This was an important finding, as two different data sets, each with its own subtle bias, pointed to an ecological disturbance of epic proportions just after the Triassic-Jurassic boundary.

Marlene reconstructed the magnificent arching ferns of *Thaumatopteris* found in the Disaster Bed by welding models of its Triassic fronds using copper foil, as they would have been too fragile made from paper. She positioned the complete models in a variety of lifelike poses and drew from them. Creating models of both plants and animals allowed her to reconstruct them in an accurate, lifelike way. Marlene's studio became the Jurassic landscape of the Disaster Bed during this intense phase of artwork, her computer screen peering out from the undergrowth of copper-bodied fern foliage.

FIGURE 3.10. Marlene Hill Donnelly, *Monotony: A Portal into the Earliest Jurassic*; *facing:* Marlene Hill Donnelly, *Modern Fern Fiddle Heads*. Pencil drawing.

MONOTONY

While the extinction event led to opportunity for the ferns and horsetails, it also created a much more monotonous landscape dominated by this "recovery" vegetation. We gauged this change in the landscape from heterogeneous Triassic forests to more monotonous fern marshes by studying the plant fossil abundance data. Statistical analyses were used to quantify how different the species composition of each of our individual excavation pits was from those of the other three collectors' pits in the same bed. In the Disaster Bed, laid down when atmospheric carbon dioxide and sulfur dioxide concentrations were high, plant diversity remained low. Conditions favored those taxa that could thrive in disturbed environments.

At this time before the evolution of flowering plants, this warmer and seasonally much wetter environment benefited ferns. Like horsetails, ferns are capable of reproducing vegetatively (without sex) by means of underground stems called rhizomes, which can sprout whole new plants. Rhizomatous growth is well documented in modern *Dipteris* and probably occurred in its ancestors, *Thaumatopteris* and *Dictyophyllum*, which were abundant in the Disaster Bed. In addition, energy storage in rhizomes affords a backup strategy for ferns, allowing them to hang on through tough times. Such vegetative reproduction leads to dense stands of vegetation made up almost exclusively of the same species, sometimes of the same individual plant, which clones itself again and again across whole landscapes. In some instances, ferns can reproduce asexually by a process called apogamy ("virgin birth"), in which an embryo is produced without fertilization of sperm and egg. In essence, this is another form of genetic cloning, like the cloning that produced Dolly the sheep, the first genetically cloned mammal; however, it is almost ubiquitous in ferns. It is certainly a handy trait to have in the midst of a disaster when populations have dwindled and only a few females of a species remain.

WILDFIRE

Smoke from wildfires billowed across the Greenland sky in the Early Jurassic. While fire cannot be preserved directly, we can piece together its history in deep time through the identification of indicators or proxies. These proxies include fire scars on fossilized trees, the presence of polycyclic aromatic hydrocarbons (ring-shaped molecules uniquely produced by burning) in sediments, and the presence and abundance of fossilized charcoal. Charcoal is perhaps the most widely used proxy for fire in deep time. It can be recognized by its color, luster, and microscopic properties. Plant parts of the past that have burned and charcoalified are usually exquisitely preserved. Their shiny blackened forms can reveal in perfect cellular detail all their internal workings—how and where water was moved around the plant when it was living, how tendrils reached around the stems of other plants, and where fibrous tissues occurred in the plant body to provide structural support. Charcoal has been used to document Earth's wildfire history as far back in the fossil record as 430 million years ago.

Throughout our studies up and down the cliffs of Astartekløft, we were constantly on the lookout for fossilized charcoal. We became skilled at spotting little black shiny jewels of charcoal among the grays of the sediment. The record of shifting charcoal abundance that resulted from our collecting efforts shows a fivefold increase in wildfire activity across the Triassic-

FIGURE 3.11. Marlene Hill Donnelly, *Wildfire: A Portal into the Earliest Jurassic.*

FIGURE 3.12. Marlene Hill Donnelly, *Volcanic plume study*. Watercolor and ink.

Jurassic boundary. But why was there such a dramatic increase in wildfires in Greenland in the earliest Jurassic? It is almost certainly because there was an increase in fire susceptibility linked to the changes in climate. The jump in atmospheric carbon dioxide concentrations from the latest Triassic to the earliest Jurassic led to profound greenhouse warming and a change in plant life. Both climate models and modern observations inform us that rising carbon dioxide concentrations lead to higher global temperatures. This warming, in turn, leads to increased variation in the water vapor content of the troposphere (the layer of the atmosphere closest to Earth's surface). A warmer and wetter troposphere drives more storms, and the result is that lightning activity increases sharply. Lightning strikes today, as in the past, are the main starter of wildfires. The recent 1°C (1.8°F) increase in global average surface temperature has been linked to a 40 percent increase in lightning activity! The atmospheric carbon dioxide rise across the Triassic-Jurassic boundary would have resulted in a rise in average global annual temperature to 7°C (12.6°F) above the modern average, though locally in Greenland, it would have been far greater. This global warming almost certainly resulted not only in more lightning strikes, but also in more rapid drying of plant fuels, which would have made plants more susceptible to ignition and combustion. All of these factors mean more wildfires.

Surprisingly, however, increased storminess and lightning strikes may not have been the main cause of the smoke-filled Jurassic skies. Our own studies have shown that changes in vegetation structure and architecture were the root cause of the dramatic increase in the number of wildfires. Plants and plant parts are fuel for fires. The fossil plant specimens from Astartekløft and other sites across East Greenland showed a pattern of decreasing leaf dimensions across the Triassic-Jurassic boundary. The new species that arose in the earliest Jurassic had almost universally narrower leaves. This was probably an adaptation to hotter climates, as narrow, highly divided leaves lose heat via convective cooling more effectively than broad leaves. Fire calorimetry experiments by Claire Belcher, at the University of Exeter, on modern plants have shown that narrow-leaved plants are typically more flammable than those with broad leaves. The observation of this phenomenon in the plant fossil record provides a sobering example of the interconnectedness within the Earth system. It demonstrates how biological adaptation to climate change can have knock-on consequences for fire ecology. It also serves as a warning of some of the likely consequences of human-driven global change.

FIGURE 3.13. Marlene Hill Donnelly, *Cladophlebis*. Pencil drawing; *facing:* Marlene Hill Donnelly, *Disturbance: A Portal into the Earliest Jurassic.*

DISTURBANCE

All the data gathered from our sediment and fossil collections from the Disaster Bed pointed to disturbed ecosystems. The proportion of woody species was at an all-time low. The sediments were black with charcoal, and further analysis confirmed our hunch from the field that wildfires were more frequent in the earliest Jurassic than at any time recorded in the Late Triassic beds. We feel that the prevalence of fire would have played a large role in maintaining disturbance in Early Jurassic terrestrial ecosystems. Research on modern tropical forests has shown that once a region has been burned, it becomes more susceptible to being burned again, and that the subsequent fires are often far more intense and damaging. Fires in these ecosystems change

the canopy structure and even the local hydrology, rendering them drier and more flammable.

We used the concept of singletons to test our hypothesis that the Early Jurassic landscapes were indeed anything but stable. *Singletons* are single instances in a fossil record where a new species appears as a fossil and then seemingly disappears again. We found a much higher proportion of singletons in all the Jurassic beds than in all the Triassic beds. In the Triassic, fossil species had much longer records, occurring again and again in subsequent beds. Together with the high abundance of charcoal, this finding supported the idea of ecological disturbance.

Not all fires are equal. Depending on their type, intensity, duration, and frequency, fires will affect an ecosystem to quite different degrees. Surface fires that combust the litter layer, herbs, small shrubs, saplings, and trees are the most common. Their severity is extremely variable. A high-severity fire can kill most, if not all, the trees it encounters, while a low-severity fire may result in little to no mature tree mortality. When a surface fire has sufficient ladder fuel, however, it can climb upward from the surface through branches and into the canopy of the forest. Then, if driven by wind, steep topography, and close vegetation growth, it can burn as a crown fire—one of the most intense types of fires. An actively spreading crown fire is almost always catastrophic for the forests that it rips through. We have seen these types of fires recently across Australia and California, and indeed, the reddened, ominous skies they produced were used by Marlene as analogs for those that may have occurred in the Early Jurassic when fires raged across Earth.

DIMORPHODON

In our reconstruction of the earliest Jurassic landscape, a flying reptile called a pterosaur flees in terror from the fires on the far shore. This pterosaur, *Dimorphodon*, was reconstructed from a specimen unearthed in early Jurassic-aged sediments from southern England. No animal fossils were uncovered from the sediments of Astartekløft, with the exception of one small, unidentifiable fish bone fragment. Here, we have used artistic license, in the knowl-

FIGURE 3.14. Marlene Hill Donnelly, *Dimorphodon: A Portal into the Earliest Jurassic.*

edge that a related genus, *Eudimorphodon*, was found north of Astartekløft in slightly older sediments of the Fleming Fjord Formation. *Eudimorphodon* is a genus that, to date, has been described only from the Late Triassic and not from the Jurassic, so the flying beast in our portal view could be regarded as a close cousin of the last survivors of the *Eudimorphodon* species.

The anatomy of *Dimorphodon* indicates that it was not the greatest long-distance flier, as it was heavy bodied and may have been adapted to hunting on the ground. However, it was capable of flight. Given the proximity of the United Kingdom to Greenland at the time (within 350 km) and the prevailing geography between them, without geographic obstacles such as large seaways or high mountains, it seems reasonable to deduce that *Dimorphodon* may have hunted through what remained of the forests, or plied the skies, of Jameson Land much as *Eudimorphodon* did before it.

Marlene's illustrations of *Dimorphodon* were informed by models in much the same way as her illustrations of plants. Beginning with fossil skeletons, she used measurements to construct a wire skeleton. Then she sculpted the skull (and other rigid elements) over the wire with polymer clay that hardens when baked. Finally, muscles and flesh, sculpted from non-drying clay, went on. This method gave her a fully movable "action figure" model that she maneuvered into different positions. Interestingly, according to Marlene, clay stretches in much the same way as real muscles, giving a very natural look.

VOLCANISM

We had a long scientific and artistic discussion on how best to represent the main driver of environmental upheaval and biological crisis in Triassic-Jurassic time. Although the scientific community concurs that a massive flux of greenhouse and sulfur-rich gases from the CAMP volcanic episode was that main driver, the geological evidence placed that great outpouring of lava and gases thousands of miles away from East Greenland. In terms of area, the volcanic episode formed one of the most extensive flood basalts on Earth. It is estimated to have produced enough lava to completely bury either, at the low end, Mexico or at the high end, India, under 1 km of lava. Despite these impressive statistics, it would not have been visible from Greenland, even on a good day. Our approach was to incorporate heavy hints at the existence of this massive geological force through its likely effect on the color of the Greenland skies.

Today and historically, volcanic eruptions have been associated with the phenomenon of blood-red skies. The Laki eruption in Iceland in 1783–1784 is said to have resulted in a "famine haze," whose devastating effects on live-

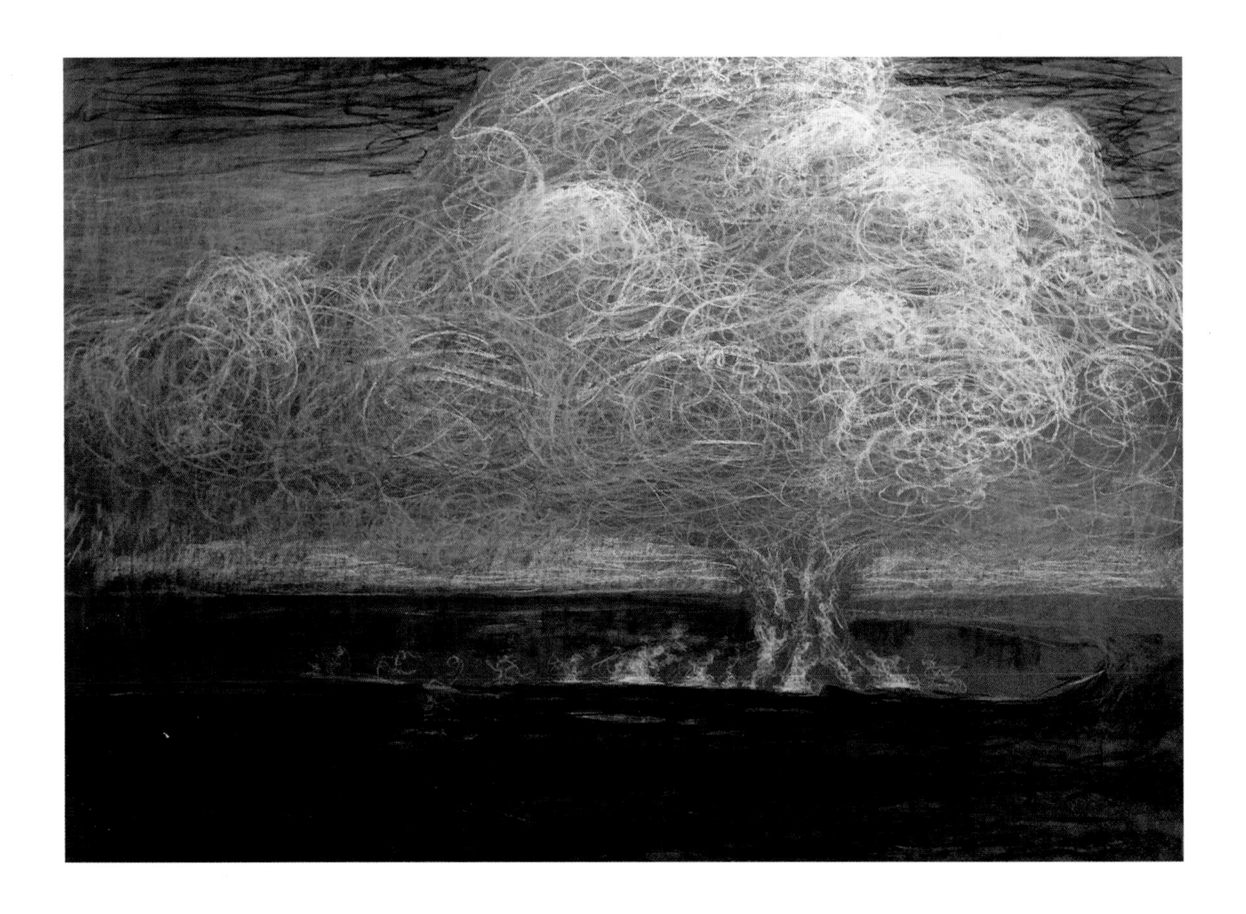

FIGURE 3.15. Marlene Hill Donnelly, *Halema'uma'u Cloud*. Mixed media.

stock and crops resulted in the deaths of over 10,000 people in Iceland. Its effect was witnessed across the entire Northern Hemisphere, where dry sulfuric acid fogs and red suns were recorded in historical archives. Particles of dust and sulfuric acid are released into the stratosphere (the layer of the atmosphere above the troposphere) by volcanic eruptions like Laki and CAMP. If they are not rained out, they remain suspended in the stratosphere. These aerosols are invisible during the day, as they are obscured by the blue of the sky. However, once the sun no longer illuminates the troposphere, but still lights the stratosphere, incredible twilight red and orange afterglows result. Changes in plant leaf shape and cuticle surface structure strongly indicate the presence of CAMP-derived sulfur dioxide aerosols in the atmosphere at the time the Disaster Bed was deposited in the rock record. Atmospheric models also propose relatively high sulfur dioxide concentrations globally. The red skies of our Disaster Bed reconstruction pay homage to these multiple lines of evidence linking both the timing and the environmental impact of volcanism to the mass extinction.

FIGURE 3.16. Marlene Hill Donnelly, *Volcano 1.* Mixed media.

Research aiming to date the CAMP volcanism has shown that successive lava flows occurred in four pulses or phases spanning a roughly 600,000-year period. These pulses correlate well in age with the Triassic-Jurassic mass extinction event, but also overlap the subsequent recovery phase, captured at Astartekløft by the Disaster Bed. Every pulse of volcanic activity would have acted like an adrenaline shot to the global atmosphere, shocking the system and throwing volcanic particles and gases high into the stratosphere.

FIGURE 3.17. Marlene Hill Donnelly, *Disturbed Arctic: A Portal into the Earliest Jurassic Landscape of East Greenland.*

4 Recovery of a Tropical Arctic

Studies of Earth's infamous mass extinctions suggest that full biological recovery of animal diversity took millions and millions of years. One of the most unexpected findings from our expeditions to East Greenland was the rate at which plant biodiversity rebounded after 85 percent of plant species went extinct across the Triassic-Jurassic boundary. We expected that recovery would take millions of years, as it did for animals, but observed it within hundreds of thousands of years. This was a heartening, but still sobering, discovery in the context of our current biodiversity crisis, which has been brought about by unsustainable human development, overharvesting, land use change, and general exploitation of wild spaces and species. Climate change is now putting additional pressure on our global biodiversity.

Our research, and that of many others, has shown that the vegetation of this Earth shows extraordinary resilience in its ability to withstand the most extreme geological forces thrown at it. These forces include meteorite impacts and volcanic events, both of which have exerted planetary-scale climate and environmental effects, including rapid global warming, global cooling, hemisphere-scale acid rain, and raging wildfires. This is a macroscale view, however, like looking at the detail of the moon with binoculars. We have discovered, by digging down through the layers of sediment at Astartekløft, that plant biodiversity alone is not a good measure of the functioning of past ecosystems. Bernie Krause, an ecologist who uses sound to study modern landscapes, remarked on a similar phenomenon in areas restored after

FIGURE 4.1. Marlene Hill Donnelly, *Delta outlet sketch*. Mixed media.

logging. They looked the same as before, but the absence of their once rich soundscapes revealed the loss of unseen species and the functional changes in the ecosystems as a whole. If only we could sound-record past landscapes. In our final, and perhaps most challenging, landscape reconstruction, we try to visually convey the scientific message of a functionally changed time and place. The Jurassic ecosystems and plant life of East Greenland had indeed recovered in terms of biodiversity, biomass, lushness, and greenness, but they were certainly not restored mirror images of the Triassic forests. They were very different ecological beasts, with measurable differences in species composition and functioning. The recovery, though rapid in geological terms, took the equivalent of over 2,500 human generations, or, thinking about it another way, about the same amount of time that elapsed between the middle Paleolithic, when Neanderthals roamed Earth along with early modern humans, and the present day.

RECOVERY

We hand-picked small black fragments of fossilized leaf cuticle directly from the fossils excavated from Astartekløft. This was a fiddly job with tweezers, but each 200 million-year-old fragment peeled from the rock with more ease than a price sticker from a gift. The blackened fragments glowed acid green when placed under a UV light source, demonstrating a phenomenon called autofluorescence. The cuticles of all leaves, both living and fossilized, have this unusual property because UV light is absorbed by phenolic acids and flavonoids within the cuticle and re-emitted in longer visible wavelengths, such as the acid green we observed under a fluorescence microscope. A simpler analogy, and one that may be more familiar, is a white T-shirt that glows blue under a UV light source at a party. In daylight, we see the T-shirt as white, but if we could see in the UV range, as bees do, we would see the T-shirt as a fluorescent blue, as it appears under a UV bulb. Fluorescence microscopy therefore allows scientists to see fossil leaf surfaces as bees see flowers—it makes the black, impenetrable fossilized leaf surface visible in cellular detail. This almost magical phenomenon allowed us to observe and record the tiny hairs, glands, cell walls, and microscopic stomata on our fossil leaves.

As shown throughout this book, plants are remarkable monitors of the climate conditions in which they grow. Take, for example, the number of stomata on the leaf surface: the higher the concentrations of carbon dioxide in the atmosphere, the fewer stomata develop on a leaf. Thus, by counting stomata on fossil leaf surfaces under the fluorescence microscope, we were able to quantify the rise in atmospheric carbon dioxide (CO_2) across the Triassic-Jurassic boundary. Peak carbon dioxide concentrations occurred in the Boundary Bed and the Disaster Bed, followed by a drop to more normal levels in the *Spectabilis* and Ledge Beds, the youngest Jurassic beds preserved at the top of the Astartekløft cliffs. While some Triassic plant species went extinct, much of the terrestrial vegetation recovered as soon as the lofty carbon dioxide concentrations dropped to more reasonable values of around 600 ppm. All measures of ecosystem health, complexity, and diversity indi-

FIGURE 4.2. Mark Widhalm, *Fragments of 200 Million-Year-Old Fossil Leaves from East Greenland*. Studio photograph. The Field Museum of Natural History.

cated a strong rebound in the *Spectabilis* and Ledge Beds. This ecological recovery is apparent in our first portal view of the Jurassic landscape. Although animal life in the oceans remained heavily afflicted, particularly the corals and reef-building organisms, life on land bounced back spectacularly.

As we look through this and later portals into the Jurassic landscape, a variety of familiar plants are visible. Some were present in the earliest Jurassic in very low abundance, most notably *Czekanowskia*, and we see the broad, arching fern *Dictyophyllum*. Others, such as *Ginkgoites*, were noticeably absent during the peak of the environmental crisis, but have reappeared. When studying fossils through geological time, we generally record their occurrences as *x*'s on a geological chart. These *x*'s note the times and places where particular fossil species have been found in the rock strata. Where there is no *x*, does this mean that the species went extinct? Perhaps the species just migrated somewhere else. Or perhaps an absence or an interruption in a fossil species' record through time just means that we scientists have not looked hard enough to find it. Alternatively, maybe it *was* present in the living landscape but never preserved as a fossil. These are the usual conundrums that we must face while working with fossils to bring the past to life.

Fossil species with apparent gaps in their geological record (indicated by an absence of *x*'s on their chart) are known as Lazarus taxa, a term first coined by paleontologists Karl Flessa and David Jablonski in the 1980s. Lazarus taxa are fossil species that appear to have gone extinct in a particular place and time, only to reappear much later—as if they are back from the dead. *Podozamites* could be referred to as a Lazarus taxon. It was abundant everywhere across the Greenland landscapes in the Triassic, like oaks across California or maples in Japan. Its sudden disappearance from the Disaster and *Spectabilis* Beds suggested that we had recorded an extinction of this magnificent, towering broad-leaved conifer. To our joy, it reappeared in the Ledge Bed right at the top of the Astartekløft cliffs. It is featured later in the full Jurassic landscape reconstruction as a rare species. Can you find it? Clearly, *Podozamites* did not become globally extinct at the Triassic-Jurassic boundary. As local environmental conditions became less favorable for the growth and survival of *Podozamites* in the Astartekløft region, this species must have

FIGURE 4.3. *Facing:* Marlene Hill Donnelly, *Recovery: A Portal into the Jurassic*; Marlene Hill Donnelly, *Dictyophyllum*. Photoshop model.

migrated to another geographic location where conditions were more favorable. Where exactly this area was is uncertain. Regions or areas that protect species diversity in this way are called refugia, and they probably occurred at latitudes north of the Greenland study sites, where the climate was cooler.

The sudden reappearance of *Podozamites* in the Ledge Bed after its apparent absence for hundreds of thousands of years indicates that climate and general environmental conditions in the region must have improved. The lesson in this example is that migration is a mechanism by which species can withstand global climate change and avoid extinction, as long as the pace of climate change is sufficiently slow to enable species to move and find new habitats and biogeographic ranges. Today, there are many examples of species' biogeographic ranges that have changed due to climate change—particularly on mountaintops and in tundra landscapes. Tree lines on mountains across the world, which mark the upper elevation limit of tree species, have been shifting upslope with climate change, and for many tropical mountaintop species, such as the Peruvian russet-crowned warbler, there is simply nowhere else to go. They have reached the upper limits of their potential biogeographic ranges, and local extinctions, called extirpations, are now taking place in their populations lower down the mountain, where temperatures are simply too hot for them to survive. Plants, of course, do not have legs or wings, and can move only as populations via dispersal of their seeds or spores, not as individuals. Today, barriers such as roads, cities, and agricultural landscapes stop the spread of certain species; so, too, does the ferocious pace of climate change.

FIGURE 4.4. Marlene Hill Donnelly, *Shape-Shifting: A Portal into the Jurassic.*

FIGURE 4.5. *Facing:* Marlene Hill Donnelly, *Sketches of Australian she oak* (*Casuarina*).

SHAPE-SHIFTING

Examination of trees in modern ecosystems shows that leaves high in the canopy, which are exposed to more sunlight, tend to be smaller than those lower down on the plant, which are more shaded. They also tend to be more lobed (divided) and to have more complex edges. These adaptations allow the leaves to dissipate absorbed heat more rapidly. Smaller leaf parts form thinner boundary layers of still air on the leaf surface. This means that water loss from the leaf is less impeded. We have observed that deeply divided or dissected leaves are often associated with warm climates, which would have existed in the Early Jurassic. It has also been noted by scientists studying *Ginkgo* that deep divisions in leaves may be related to both high sunlight exposure and high moisture availability. However, in studies of flowering plants, it has been suggested that deep leaf divisions may correlate with aridity as a way of reducing the stress of transporting water laterally within a leaf.

The fossils we uncovered in the topmost beds of the cliffs at Astartekløft—the *Spectabilis* and Ledge Beds—were truly spectacular in their varied forms. They were almost black, thick, and peeling off the rock surface. Nearly all shared an overall leaf architecture featuring highly divided leaf blades with elongate and narrow dissections. An extinct plant group called *Czekanowskia* was the most extreme in its lengthwise distortion of a typical leaf, looking superficially almost like Australian pine, also known as "she oak" (*Casuarina*), or a greatly distended *Ginkgo*, to which it may be related. Although it can be difficult to infer climate signals from the shapes of really ancient leaves, it has been shown through burn experiments on modern plants that highly dissected leaves like those literally falling out of the

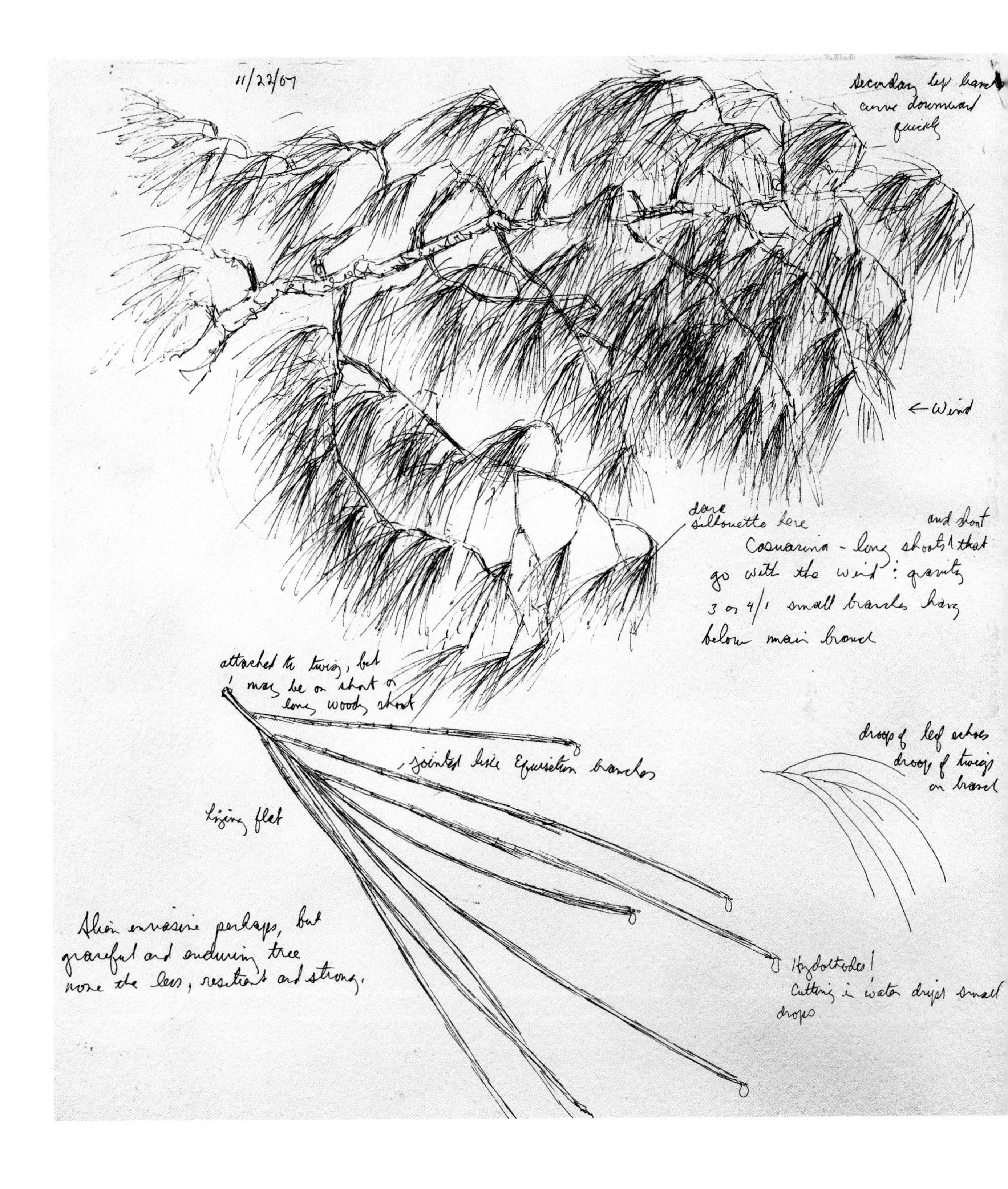

11/22/07
Secondary leaf branch
curve downward
quickly
← Wind
dark
silhouette here
Casuarina – long and short shoots that
go with the wind: gravity
3 or 4/1 small branches hang
below main branch
attached to twig, but
may be on short or
long woody shoot
jointed like Equisetum branches
lying flat
droop of leaf echoes
droop of twigs
on branch
Alien invasive perhaps, but
graceful and enduring tree
none the less, resistant and strong.
Hydathodes!
Cutting in water drips small
drops

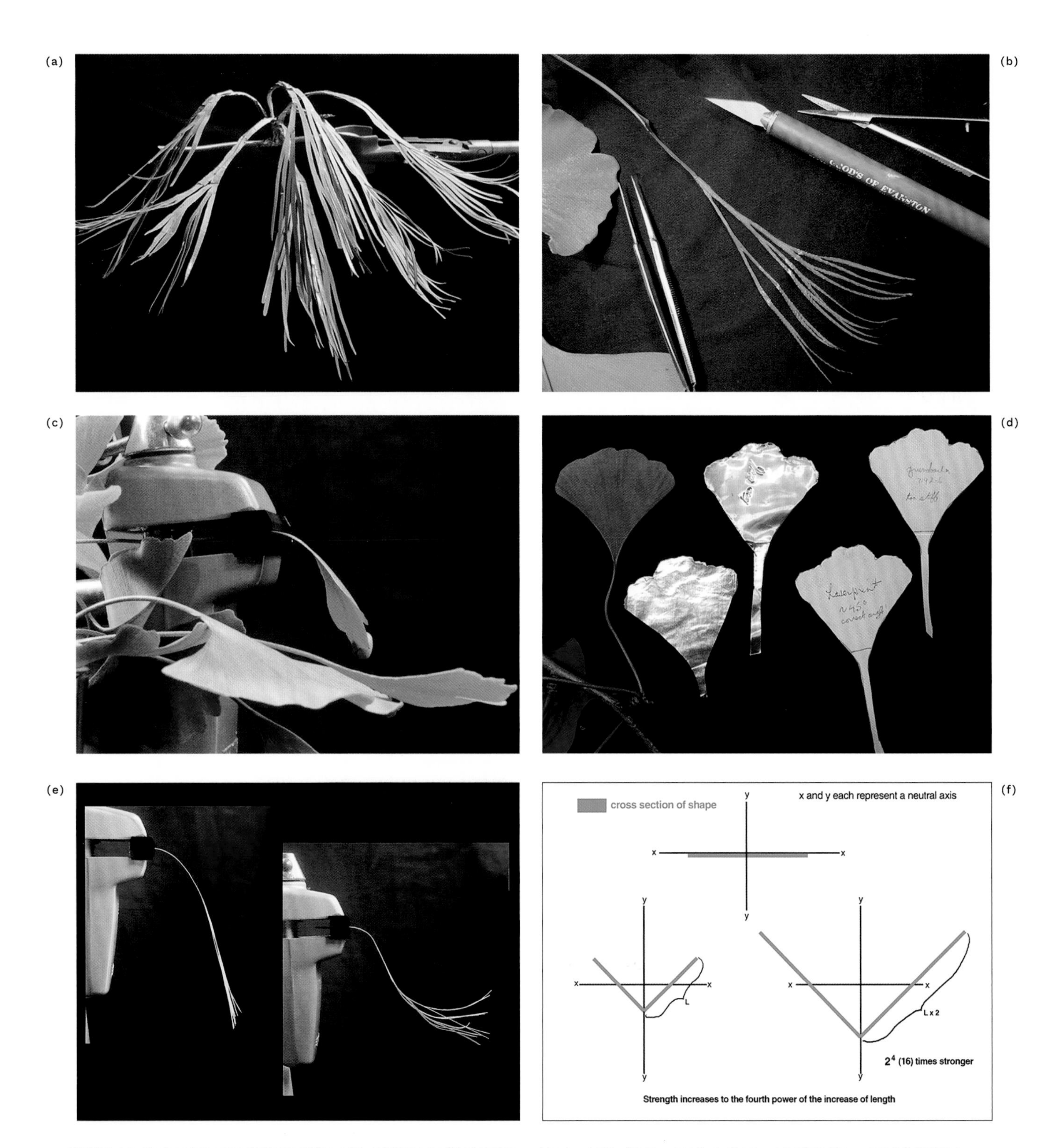

FIGURE 4.6. Engineering a new *Czekanowskia* model. (a) Paper model of *Czekanowskia* showing leaf form when "veins" are scored into the paper; (b) *Czekanowskia* model leaf cut out of modern *Ginkgo* leaf tissue; (c) demonstration of the form (habit) of a *Ginkgo biloba* leaf when held in a jewelers vise; (d) a comparison of model *Ginkgoites* leaves made from different materials (tin, copper, and copier paper of different weights) compared with a real *Ginkgo biloba* leaf in green; (e) a comparison of *Czekanowskia* paper leaf models when scored without (*left*) and with (*right*) leaf veins; (f) engineering principles behind leaf veins.

rock strata of the *Spectabilis* and Ledge Beds would have been much more flammable than their broader-leaved ancestors in the forests of the Triassic.

Bringing *Czekanowskia*, this most extreme of fossil plant forms, to life and placing reconstructed individuals of this remarkable species in the landscape were huge challenges. Again, it was art, rather than science, that came up with the pertinent questions *and*, arguably, the answers, with a little help from engineering. We knew the size, shape, and probable attachment of *Czekanowskia* leaves from the fossil evidence. But what was the most likely life pose of these leaves? Would an unusually long leaf be able to stand tall, or would it sag under its own weight?

First, with scissors and glue, Marlene made a *Czekanowskia* leaf out of living *Ginkgo* leaves. This gave a feel for how the leaf would look (delicate!), but no information regarding its life pose. To determine reasonable possibilities that didn't defy the laws of physics, Marlene consulted with mechanical engineers and ran two simple experiments; without these, she felt we would just be making things up. The first was a "modulus of elasticity" experiment to determine what available test material would be most similar to the live leaf. Next, a "second moment of area" experiment was performed in Marlene's studio. This test measures the strength of a cross-sectional shape to resist bending under its own weight. In other words, if veins were scored into the leaf surface, changing the leaf shape, how would this alter the overall pose of the reconstructed *Czekanowskia* leaf? Marlene became the scientist. She tested a variety of thin materials that seemed as similar as possible to the leaf thickness of *Czekanowskia* (we used *Ginkgo* as a model, together with our own estimates from the fossils): copper foil, plastic, and various papers. A live *Ginkgo* leaf was the control sample for *Czekanowskia* and *Sphenobaiera*, which both went through this rigorous engineering process.

Each test material was cut into the life-sized leaf shape drawn from the fossil and suspended from a jeweler's vise at the same angle. The material that bent to the position most similar to that of the real living *Ginkgo* leaf won the award for having the modulus of elasticity closest to that of the real leaf. Oddly but conveniently, the winner was plain copier paper. As we all know from making paper airplanes, the way a thin material is scored or folded drastically affects its ability to hold a shape. To find out how much stress any given leaf shape could actually take, we used the art of origami. Try this quick second moment of area origami experiment yourself: Hold a piece of paper by the edge—it will collapse limply. Then fold it down the middle and hold it up by one end at the crease—this time, it will hold a much more rigid V shape. Marlene scored "vein" creases, based on her observations of the veins in the actual leaf fossils, with a ballpoint pen into the flat paper leaf shapes.

The paper model of a slender, delicate, complicated *Czekanowskia* leaf had no body whatsoever by itself—it just flopped. But when Marlene scored veins into it, the model leaf lifted and took on a natural, sinuous shape. This ingenious experiment also showed that the leaves could not possibly have stood stiffly upright, as they had in a previous reconstruction. A new *Czekanowskia* model was born.

GENETICS

Today, there are over 350,000 different species of flowering plants, yet genetically, they are all more similar to one another than to living *Ginkgo*, which is a gymnosperm. Gymnosperms, including conifers, are seed plants whose seeds (Greek: *sperma*) are "naked" (Greek: *gymnos*), unlike those of flowering plants, which are enclosed by an ovary. So, ginkgos are more closely related to living conifers than to flowering plants. Our view of the Jurassic landscape illustrates two Mesozoic relatives of living *Ginkgo*, *Ginkgoites minuta* and *Sphenobaiera spectabilis*, as well as the lacy-leaved and newly remodeled *Czekanowskia*. These species were the new dominants of the forests that became established in East Greenland following the dramatic collapse of woody vegetation at the close of the Triassic.

Individual *Ginkgo* trees are either male or female, each passing on a set of chromosomes to the next generation. This type of reproductive setup is called dioecy (from the Greek: "two households"). The majority of plants and animals, including humans, are diploid: that is, they have two paired sets of chromosomes in every cell of the body, one set from the father and one set from the mother. Sometimes, however, a duplication of this genetic blueprint can occur in a process known as whole-genome duplication. This process is usually fatal to humans, as it interrupts every aspect of healthy development.

FIGURE 4.7. Marlene Hill Donnelly, *Genetics: A Portal into the Jurassic*; *facing:* Stephen Hesselbo, *Sphenobaiera spectabilis*. Field photograph. Stephen's hand for scale.

But it can be highly advantageous in many other organisms, especially plants, as it provides an extra set of genetic material on which natural selection can act. Extra complete sets of chromosomes offer backup if mutations occur in genes, and they give plants more vigor, so that they grow faster and bigger. Individuals that have more than the usual two sets of chromosomes are called polyploids, and they are often associated with a greater overall genome size.

There has been some excitement over our studies of *Sphenobaiera spectabilis*, visible in our portal view of the Jurassic canopy, because we think it may be an ancient polyploid. We used stomatal size, which is strongly related to genome size, as an indicator of the genome size of *Sphenobaiera*. The species has giant cells and giant stomata compared with those of its modern relative, *Ginkgo biloba*. Based on this simple observation, we think it likely that *Sphenobaiera* had 1.5 to 2 times as much chromosomal DNA in each cell as *Ginkgo biloba*. If you have lots of DNA, you need a big nucleus to accommodate it, and a big nucleus can only be contained in a big plant cell!

Our study is at a very early stage and requires further verification, but if our observation can be supported with further data, it would confirm the hypothesis that polyploids have an ecological advantage in the aftermath of environmental disaster. This hypothesis stems from the idea that polyploids have double the range of genetic variation found in diploid plants. In other words, they have more tools in their toolbox to choose from to survive an environmental crisis. Many researchers believe that whole-genome duplication is critical for the formation of new species and as a mechanism to increase plant biodiversity. Polyploids are often differently shaped and reproductively isolated from their parents. These differences are difficult to convey visually without adding interpretive text that draws attention to the giant size of the *Sphenobaiera* leaves compared with those of their relatives mingled among them high in the canopy of the Jurassic forests. The leaf sizes, shapes, vein densities, and comparative thicknesses of all the *Ginkgo* species from Greenland show phenomenal variation, particularly in the Jurassic. It seems that following the extinction event, a great burst of new species arose. It took an exceedingly long time, however—most likely on the order of hundreds of thousands of years.

CARBON

In the Jurassic, you can't "see the trees for the wood." The vegetation has shifted away from fern meadows to dense woodlands once again, indicating recovery. A similar transition can be seen today on Mount Etna, Europe's most active volcano. Recent lava flows and ash falls are bare, with only a few patches of herbaceous plants such as Sicilian milk vetch (*Astragalus siculus*) and Sicilian soapwort (*Saponaria sicula*) dotted here and there. Old lava flows on the mountain, dating back tens of thousands of years, are fully forested with diverse stands of oak and chestnut because the vegetation on these old flows is fully recovered. The resurgence of forests in the Jurassic took hundreds of thousands of years. Once recovered, these forests were a tremendous carbon sink, locking up large quantities of the carbon dioxide released during the great outpouring of lava that would eventually become the basalt of the Central Atlantic Magmatic Province.

All living organisms are formed, at least in part, of compounds that contain carbon. The movement of carbon among the Earth system's various components, including the atmosphere, rocks and sediments, oceans, lakes, and rivers, and of course, living organisms, constitutes the carbon cycle. The role of plants in the carbon cycle is critical, as they absorb carbon dioxide and then, via the two-stage process of photosynthesis, convert it to sugars. The first stage uses chlorophyll to trap the sun's light energy and convert it to chemical energy in the form of ATP (adenosine triphosphate). The second stage uses this chemical energy to convert water and carbon dioxide to sugars, fixing the carbon and in the process releasing oxygen back to the atmosphere. Photosynthetic organisms are known as primary producers. Without primary producers, there is no ecological stability, as all consumers (such as animals) are reliant on them for energy. Ecological modeling has shown that

FIGURE 4.8. Marlene Hill Donnelly, *Carbon: A Portal into the Jurassic.*

entire food webs can collapse if the base of a food web—the plants—is disturbed. Just like dominoes, every tile comes crashing down if the first domino is pushed. In our portal view, you can see a consumer: *Dracoraptor*, a shoreline predator or scavenger described from Wales in the UK. This vertebrate species is one of the few known Jurassic (Hettangian) theropods that would have sat at the top of the food chain. Such carnivores are highly susceptible to interruptions in the food chain below them and are liable to go extinct if their prey becomes scarce or disappears. *Dracoraptor* was one of the lucky survivors of the Triassic-Jurassic extinction event.

As animals consume plants, they break down carbohydrates as part of respiration, which releases carbon dioxide back to the atmosphere. Plants respire, too. When organisms die and decay, they release the carbon in their bodies. However, if the organisms are buried and fossilized before they can decay, their carbon is locked away. Locked away, that is, until that carbon (in the form of carbon reserves such as coal, natural gas, and petroleum) is brought back to the surface and burned, releasing energy as well as the stored carbon. The volume of biomass contained in woody plants and the complexity of the bonds in wood means that trees take longer than other plants to decay. It follows, therefore, that trees growing in swamps and mires like those that fringed Kap Stewart Lake in East Greenland were especially likely to be preserved as fossils.

This locking up of carbon—its sequestration from the atmosphere—in buried trees and other plants that eventually became coal was one of a number of critical mechanisms that prevented the Jurassic Earth from completely burning up. This process was and still is one of Earth's vital thermostats. The pace at which we are currently burning carbon reserves, in the form of fossil fuels that were formed on time scales of millions and millions of years, is a cause for great concern. The planet has not been able to respond to our re-

FIGURE 4.9. Marlene Hill Donnelly, *Human Volcano*. Abstract ink on paper.

lease of carbon as it did during the Jurassic. Humans are now a geological-scale force acting on our entire Earth system, and our actions are altering the atmosphere and climate more rapidly than CAMP volcanism did 200 million years ago. The Earth system has the same thermostat today, in the form of carbon sequestration by trees and other plant biomass, but it simply cannot keep pace with the sheer volume of greenhouse gases we are pumping into the atmosphere year after year.

ENERGY

The fossils from Greenland were all preserved in sediments eroded and then deposited by rivers, flooding events, and winds. The distance the sediment traveled, and its size, angularity, and sorting, all act as important data points for interpreting the energy and style of sediment transportation. The geological characteristics of the sediments from the *Spectabilis* and Ledge Beds indicate that they were deposited in a delta or floodplain close to Kap Stewart Lake—the ancient freshwater lake whose site we had flown over by helicopter on our way to our first field site at Ranunkeldal. Our studies of the sediments at Astartekløft revealed that the river channels feeding the lake in the earliest Jurassic carried high-velocity, erosive flows. In contrast, by the time of deposition of the *Spectabilis* and Ledge Beds, when forests had once again established themselves across the landscape, the rivers were slower and more meandering, producing abandoned channels and little oxbow lakes in a large floodplain or delta.

Deltas form when sediments carried by streams are deposited due to a sudden reduction in flow velocity. Such slowing can occur as a river enters a lake. The heaviest sediment particles drop out first, and the lightest are carried farther toward the delta front. This sedimentation process can be augmented by vegetation that further slows the water flow. When suspended clays and fine silts coagulate as the water chemistry changes, large clumps of those materials form and sink. This process often gives a characteristic cloudy hue to the water, a detail that Marlene has included in the sunlit portion of our view of the Jurassic. Eventually, the water shallows, and topset beds rise above the surface of the water, forming a new alluvial plain. As more sediment is deposited, especially by flooding, the river becomes choked and branches into multiple channels. As sediment blocks these channels, new ones branch off, and the channels shift, leaving abandoned channels, such as those where many of our fossils were collected. As these newly exposed land surfaces are colonized by plants, they slowly grow upward until the land becomes dry. Understanding the sedimentology gives us a better appreciation of these shifting past landforms. In our full reconstruction of the Jurassic landscape, the viewer is placed on top of one of the dry banks of an abandoned channel, now colonized by a dense growth of plant life, looking out across a delta at the mouth of a river feeding Kap Stewart Lake. The immersive nature of this ancient vista comes from Marlene's capacity to transport us from her field sketch locations into the past. This time travel seems possible only when the languages of science and art collide.

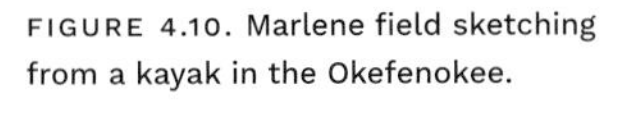

FIGURE 4.10. Marlene field sketching from a kayak in the Okefenokee.

Marlene used the sedimentology data on past landforms to determine what modern locations were appropriate for immersive field sketching and journaling to get a feel for what happens in a similar living environment. What you see in our reconstruction of the Jurassic landscape, as with the Triassic landscape previously, is a fusion of scientific data from the fossils translated through Marlene's keen observations—sketched in the rain in Hawaiian rainforests and volcanic calderas, and from a kayak in the Okefenokee Swamp and the Skagit River delta. Extended field sketching was the best way for her to absorb the myriad sensations and clues that every environment contains, including the subtle information that most of us are blind to—that "feeling" you get from a place.

FUNCTION

The fossil record of plants and animals shows that extinction is normal. At any point in time, species are undergoing extinction. What is unusual about mass extinctions like the Triassic-Jurassic event and our contemporary biodiversity crisis is that the rates and magnitude of extinction far exceed what is considered normal. The rate and magnitude of the animal extinctions at the Triassic-Jurassic boundary were so severe and so far beyond normal that they have been classified as a mass extinction event. As we have seen in our

FIGURE 4.11. Marlene Hill Donnelly, *Okefenokee sketch*. Pen and ink on paper.

FIGURE 4.12. Marlene Hill Donnelly, *Delta sketch*. Mixed media.

journey through the reconstructed forests of East Greenland, although plant life underwent species extinctions, local emigration, and intense ecological upheaval, all but one plant family survived. Technically, therefore, this profound ecological crisis is not classified as a floral mass extinction. To satisfy the criteria for a mass extinction, a high proportion of *all* plant families would have to have been wiped out across the globe. But are we using the wrong metrics to classify biological crises in the geological past? With this question in mind, our team sought to consider additional ways of measuring ecosystem health and functioning and how it changed across the Triassic-Jurassic boundary.

We used the pie charts of fossil plant abundance that we generated for every fossil bed to investigate whether the size of the slice of pie occupied by each species differed between the Triassic and Jurassic forests. The pattern of change we observed was astounding. Although the four most dominant plant species in the Triassic forests survived, they became exceptionally rare in the Jurassic. The magnitude of the ecological change was equivalent to walking through an oak-holly-bracken forest today and encountering one of those species in 80 out of every 100 steps, then crossing a boundary and encountering one of the species in only 2 out of every 100 steps. The plants dominant in the Triassic included the fern *Dictyophyllum*, the broad-leaved conifer *Podozamites*, and two extinct proto-flowered genera, *Anomozamites* and *Pterophyllum*. In the Jurassic, *Ginkgoites* and *Czekanowskia* became the new forest kings. Yet both *Ginkgoites* and *Czekanowskia* were relatively rare in the Triassic.

FIGURE 4.13. Marlene Hill Donnelly, *Ginkgo biloba*. Pencil drawing on paper.

We took one further step in our exploration of the changing species composition and investigated whether it altered the functioning of these ancient floras. Biophysical models that can simulate the physiology of fossil plants on the basis of their structural (such as the density of stomata and leaf veins) and chemical attributes (such as the ratio of carbon to nitrogen in their leaves) were used to estimate the photosynthetic rates of the once living plants before they became immortalized as fossils. In the modern flora, there is a general trend whereby species with higher stomatal and vein density and higher leaf nitrogen content have higher photosynthetic rates. The modeling exercise provided us with insights on how quickly the fossil species were capable of moving water, carbon, and nitrogen through plant parts—their roots, stems, and leaves. We discovered that Triassic-Jurassic greenhouse warming preferentially benefited species with slower photosynthetic rates, such as *Ginkgoites*. Such plants are considered to have an ecological strategy of tolerance—a slow but safe life strategy. Incidentally, this strategy makes *Gingko* an excellent choice for city planting today. The plants with more "live fast, die young" strategies, such as *Anomozamites* and *Pterophyllum*, suffered heavy extinction as temperatures warmed and greenhouse gas concentrations climbed.

We think that this ecological shift in plant life strategies fundamentally changed the functioning of entire ecosystems. The forests of the Jurassic would have cycled water and carbon at a slower pace than those of the Triassic; leaves would probably have decomposed at slower rates, and the nutritional value of forage and browse for animals would have been lower. These findings suggest that the total diversity of plant species in a forest is not always the best measure of how it is functioning—sometimes you have to look at other forest characteristics. If we could have measured the soundscapes before and after the mass extinction, they would have told the same story. As scientists, we have to look deeper into the data to get the full picture.

AGES AND DATES

Geologists readily bandy about events that occurred tens or even hundreds of millions of years ago—we've done it throughout this book, as we talk about the transition from the Triassic to the Jurassic 201,360,000 years ago, give or take 170,000 years! Such vast time scales are almost incomprehensible to us, given that the entire history of humans (*Homo sapiens*) spans much less than 1 million years—perhaps as little as 200,000 years—and that our ancestors began producing food and adapting our surroundings only 12,000 years ago.

The absolute date that is given above for the Triassic-Jurassic boundary is based on work published in 2014 by an international group of scientists from Europe and the United States. These scientists studied the radiometric decay of uranium to lead in volcanic ash beds in Peru to establish an age for the boundary. This dating technique takes advantage of naturally occurring elements, such as uranium, that are unstable (radioactive). Uranium decays to lead over a known time frame. In the case of uranium-235, it takes 704 million years for half of it to decay to lead. This length of time is known as its half-life. Analyzing the ratio of uranium to its decay product, lead, in a sample of a mineral such as zircon (which occurs abundantly in volcanic ash) allows us to back-calculate the absolute age of the sample. A nice analogy for this complex dating technique is the use of "sell-by" dates for food. We know that meat, eggs, and fish have different sell-by dates because different food items have different rates of decay. In other words, they have different shelf lives. If you know the rate of decay for each food, or its half-life (the time in which half of it decays), then you can inform customers of a sensible sell-by date to keep them safe. The dating of rocks with radioactive elements like uranium-235, therefore, is just a rather long (and exceptionally precise) extension of the food shelf-life concept that we are all familiar with. Unfortunately, we did not discover any volcanic ash beds in our Greenland field sites, so we had no precious zircons in which to measure uranium-to-lead ratios. So, how do we know that our Greenland fossils occur at the Triassic-Jurassic boundary? The answer lies in a technique called relative dating.

Fossils are typically found in sedimentary rocks. These rocks are deposited in layers, such that the oldest rocks are at the bottom of the sequence and the youngest at the top. Thus, a fossil found in the top layer is younger than one found in the bottom layer. Some sedimentary layers contain volcanic ash that can be dated. This permits the layers, and the fossils they contain, to be assigned an absolute age. Repeat co-occurrences of particular radiometric dates and very charismatic fossils of a particular species establish confidence in a relationship between the two. It is then possible to use fossils that have very well established date ranges as an indicator of the ages of sediments in which they occur. This type of relative dating approach is called biostratigraphy. Just as we can use the clothes and music in a film to quickly date it to the 1970s, a paleontologist can use diagnostic fossils in a sediment layer to date it to the Late Triassic. In Greenland, the Triassic-Jurassic boundary is defined both by changes in leaf fossils and by changes in fossil pollen and spores. The Late Triassic is characterized by the presence of *Lepidopteris*, a seed plant vine, and the onset of the Jurassic by its disappearance. Jurassic strata are characterized by the presence of the fern *Thaumatopteris*, which is abundant in our portal view of the Jurassic landscape, though it first appeared in the Triassic-Jurassic Boundary Bed of East Greenland.

Exquisitely detailed pollen and spore studies of the Astartekløft beds, together with chemical studies, confirm that the lowermost six beds (Horsetail, Halfway, Fox, *Ginkgo*, Flower. and Boundary Beds) in the cliffs are Late Triassic in age and that the uppermost three (Disaster, *Spectabilis*, and Ledge Beds) are earliest Jurassic in age. In addition, a particularly important and time-diagnostic pollen grain, *Cerebropollenites thiergartii*, named after its brain-like (*cerebro-*) form, along with many others, helped to correlate our field sites with the global reference locality for the Triassic-Jurassic boundary in Kuhjoch, Austria. It is important to be able to anchor our sites in time and space with other Triassic-Jurassic localities across the world because this time in Earth history provides such important insights for our future.

FIGURE 4.14. Marlene Hill Donnelly, *Ages and Dates: A Portal into the Jurassic.*

TIME TRAVELING WITH SCIENCE AND ART

Our scientific and artistic quest to reconstruct the changing landscape of East Greenland during the transition from the Triassic to the Jurassic has revealed a tropical Arctic. In the Triassic, steamy lakeside mires teemed with dragonflies, and the soft popping of mud bubbles filled the air between thickets of towering *Neocalamites*. From the statuesque forests, branches of *Podozamites* and *Ginkgoites* reached over deep, meandering stream waters where dragonflies laid their eggs on the leaves that dipped in and out. In the quiet of the forest understory, red-tinged *Dictyophyllum* ferns unfurled among the proto-flower-laden *Anomozamites* and *Pterophyllum*. Gaps in the dense forest canopy caused by tree falls brought rare flickers of light to the forest floor, where ancient and now extinct species competed to occupy the newly lit space. *Lepidopteris* vines reached through the dead-leaf skirts of *Pterophyllum*, or perhaps clung to the bark of *Ginkgoites* and *Podozamites*, to steal the light of the canopy tops. Every morning, they oozed water from tiny guttation pores, forming beaded droplets of water, like dew, on each leaf held high in the canopy. The Triassic forests were thick, dripping. They flooded regularly and to spectacular effect. Thick muds and silts were deposited by the floodwaters as the rivers burst their banks, bringing with them a new wash of nutrients. Wildfires, although relatively infrequent, burned the understory plants—their fuel—and left tiny fragments of charcoal and ash rich in phosphorus in their wake.

These landscapes were changed forever at the close of the Triassic. By the earliest Jurassic, few forest canopy species remained across East Greenland. Now the landscape was open. It was rich in ferns and horsetails and sodden wet with mires oozing muds, but the land was not bright and exposed, despite the lack of trees. An eerie red hue colored the earliest Jurassic skies. Acidic haze and wildfire smoke eclipsed the sun. The survivors of the Triassic-Jurassic mass extinction and emigration were few, but those that could tolerate the extreme environmental conditions were abundant across the landscape. They lacked their usual competitors. The slow, deep, meandering rivers of the Triassic were replaced by chaotic channels constantly switching direction and changing their routes—disturbing and uprooting plants along their frenetic paths. Wildfires raged across the landscape with a new ferocity and intensity, burning everything in their track.

At last, later in the Jurassic, the rivers again slowed. Oxbow lakes began to form across the wide floodplains of meandering rivers once again. Forests reclaimed the landscape, adding their three-dimensional wonder, and

FIGURE 4.15. Marlene Hill Donnelly, *Recovery of a Tropical Arctic: A Portal into the Lost Jurassic Landscape of East Greenland.*

within their protection, plant species diversified and prospered, some with greater exuberance and vigor than their parental lineages (for example, *Sphenobaiera spectabilis* with its enlarged genome). The Jurassic forests of East Greenland were more open than those of the Triassic. More light made it through their filigree canopy to the forest floor, which in autumn was carpeted thickly with butternut-colored leaves of *Ginkgoites*, *Sphenobaiera*, and *Czekanowskia*, which decayed more slowly than those of their forest ancestors of the Triassic. Above the forest canopy, the skies were clear blue. The acidic haze had receded as volcanism in the Central Atlantic region had quieted.

Globally, the temperature was around 3°C–4°C (5.4–7.2°F) warmer than at present when the Triassic forests flourished, then peaked at 7°C (12.6°F) higher than at present at the demise of the Triassic and dawn of the Jurassic. In East Greenland, average local annual temperatures may have risen by as much as 16°C (28.8°F) in response to the volcanic release of greenhouse gases. Today, human fossil fuel use and land use change are exerting a geological-scale force on the global climate system whose magnitude is becoming comparable with that of the 600,000 years of volcanism that triggered the ecological collapse of the Triassic flora. The main difference is that today's human endeavor and population growth have raised the carbon dioxide content of the atmosphere 10 to 100 times *faster* than CAMP volcanism did across the Triassic-Jurassic boundary.

The forests of East Greenland were tropical in character despite the fact that they occurred at latitudes well outside the modern tropical belts. These forested landscapes responded to global climate change with species extinctions, complete but temporary loss of forests, and marked alterations in the functioning of entire ecosystems. There is also a story of hope, however, because our studies of the East Greenland fossil flora also demonstrate incredible resilience, adaptation to the new normal, emigration to new home ranges, and ultimately, rebuilding. Most of the major plant family lines survived this great environmental crisis, despite the fact that many suffered heavy species losses. During the peak of environmental change, when sulfur dioxide and greenhouse gases reached their zenith, the landscapes of coastal East Greenland were mostly barren of trees, but were flooded with ferns and weedy opportunists. Wildfires raged, and floodwaters ran off the land, bringing nutrient flushes into the oceans. The trees returned, however, and given sufficient time—hundreds of thousands of years—forested ecosystems were once again restored. Strange new fossil forms told of a burst of new species arising in the rebuilt forests of the Jurassic. These forests had a distinctly different character and species makeup, yet they emerged, in evolutionary terms, from their ancestor forests of the Triassic.

The great moral of the East Greenland story is that plants, floras, and the whole biosphere *can* adapt to the most extreme global environmental change thrown at it, as long as there is not an expectation that it will look and function the same as it did before. More importantly, perhaps, resilience is possible only if the pace of climate change does not exceed the pace of adaptation, migration, and reestablishment. It is also critically imperative that avenues for migration be available for the great latitudinal mingling of biodiversity that will ensue. It is not too late to make a difference to curb greenhouse gas emissions, conserve our current climate state, and keep the global average temperature from rising by more than 2°C (28.8°F). The world's stunning and varied biodiversity is already in crisis due to direct human development, expansion of cities and agriculture, and overexploitation of our natural resources. Ongoing climate change will magnify these pressures. Our work with the fossil flora of East Greenland shows that a 4°C (7.2°F) global temperature rise causes ecological instability and a 6°C (10.8°F) global temperature rise causes complete collapse of forests. Our exploration of the science and art of past landscapes tells us loudly that it will be too late to protect the modern landscapes that we know and so love, and the species they harbor, if we wait to solve climate change until tomorrow.

Acknowledgments

The authors are greatly indebted to a number of individuals and institutions who have made this book possible. We thank the National Geographic Society (7038-01), the Comer Science and Education Foundation (#13), the Field Museum of Natural History, the Royal Society of London, the United Kingdom Natural Environment Research Council, the European Union (Marie Curie Excellence Grant MEXT-CT-2006-042531), Science Foundation Ireland (PI grant 11/PI/1103), and the European Research Council (grant ERC-279962-OXYEVOL) for funding. Special thanks to E. C. Meeker and to R. H. and P. O. Schnadig for additional funding support to JCM through the Field Museum Women's Board Field Dream Program, and to the Grainger Foundation for funding support to IJG. Andrew Scott is thanked for essential equipment donated to IJG and his far more important academic advice.

We especially thank Finn Surlyk, Stephen Hesselbo, David Sunderlin, Matthew Haworth, and Mihai Popa for their friendship, good humor, and strong stomachs while stuck with us in various tents, but also for their tireless hard work and diligence in the field, even when faced with precipitous climbs and collecting sites suitable only for a bighorn sheep. However, we would most like to thank them for their continued invaluable advice and insights.

Many thanks to the Danish Polar Centre, Constable Pynt Airbase staff for logistic support and encouragement. IJG would also like to recognize the Constable Pynt Airbase contingent of retired huskies from the Sirius Sled Dog Patrol (Slædepatruljen Sirius) for their surprise unsupervised visits to raid his pack in the field. Thanks, too, to Rebekah Hines for arranging permits and all things logistic and for seeking out and stocking our expeditions with an abundant supply of hot sauce and Parmesan cheese (we devoured an entire wheel in 1 month), as well as food apparently irresistible to huskies!

At the Field Museum, thanks are owed to all in Geology, but especially to Paul Mayer and Jack Wittry (volunteer extraordinaire) for their help in hunting down specimens and ensuring their continued curation and care. John Weinstein and Mark Widhalm are thanked for their stunning photography, and Nina Cummings for her help with photography image curation. Paul Lane was a boon, and without his expert and diligent work in scanning, printing, and general digital wizardry, this book would not have been possible.

Gary Hoyle is thanked for his extraordinary handcrafted model of *Sphenobaiera spectabilis*, which really brings meaning to the name of this plant species! We thank Peter Lang for the illustrated map of East Greenland.

We thank the Triassic-Jurassic researcher teams around the world who have published the great scientific studies on which this book is based. In particular, we would like to thank the following for their special contributions: Luke Mander, Karen Bacon, Margret Steinthorsdottir, Claire Belcher, Wuu Kuang Soh, and Charilaos Yiotis. A huge debt is due to a former student of Tom Harris, the late Bill Chaloner, mentor to both JCM and IJG and a continuing source of inspiration. We sincerely thank our editor, Joseph Calamia, our skilled copy editor, Norma Sims Roche, and the entire team at the University of Chicago Press for guiding this project to fruition. Last but not least, we thank Nora for the inspirational title and our current institutions, Trinity College Dublin (JCM), Colby College (IJG), and the Field Museum (MHD), for their continued support.

APPENDIX

A Fossil Plant Gallery

How are fossil species named and identified in paleobotany? In this appendix, we invite our readers to delve a little deeper into the science of paleobotany and the unique species we encountered in our East Greenland study, including how and by whom various fossil species were named and the basis for their being so named and categorized. Taxonomy—the naming and categorizing of species—is a minefield when it comes to fossil plants. The reason for this is twofold. First, plants are rarely fossilized whole; instead, they are usually broken into parts, or plant organs. Second, it is difficult to define what a species is when you cannot observe its breeding behavior or directly measure its genetic information. Simplifying greatly, this lack of genetic or reproductive evidence means that paleobotanists cannot use modern species concepts, so paleobotanic species are based on similarities in the fossils' anatomy and morphology instead. Essentially, an expert will examine a fossil's shape, its internal characteristics, and possibly even its chemistry to determine whether it is sufficiently similar to any known fossil species to merit being assigned to it, or sufficiently different to require that a new species be created.

To complicate things further, unlike vertebrate paleontologists, who might typically assign all of the morphologically diverse bones of a disarticulated dinosaur to one species (e.g., the wonderfully named *Gasosaurus constructus*), paleobotanists elect to name each separate plant organ individually as a *form taxon* (singular; plural *taxa*). The practice of using form taxa in paleobotany results in multiple, often very differently named, genera and species for different organs (e.g., roots, bark, cones, and bracts) from the same plant. Only once these organs are proved to occur in organic connection with one another can they be assigned to a whole-plant concept and given a new whole-species name. However, this doesn't mean that the form taxon names for the organs are no longer used when those organs are found in a dispersed state. We could use ratatouille as an analogy. Love it or hate it, most of us would recognize this recipe composed of various ingredients (tomatoes, eggplants, peppers, zucchini, etc.) and might name it as such if we saw it. However, were we to find a lone piece of eggplant sitting on the chef's cutting board, we would still call it eggplant, even though we know it can be a component of ratatouille.

In the following sections, we highlight the taxonomy of some of the most important taxa that occupied the forests and changing landscapes of Triassic and Jurassic East Greenland.

ANOMOZAMITES

Anomozamites is a fossil genus (a genus is the taxonomic level above species that unites a group of species with broadly similar characteristics, such as *Homo sapiens* and *Homo antecessor*). It was first described by French botanist Willhelm Philippe Schimper in 1870 while he was curator at the Natural History Museum in Strasbourg, and then again by the English paleobotanist Tom Harris in 1969. *Anomozamites* (from the Greek *anomo-*, "unusual" or "without"; *azaniae*, "pine cone") is the name of a genus established for a type of fossil leaf belonging to the extinct seed plant group Bennettitales. The leaves

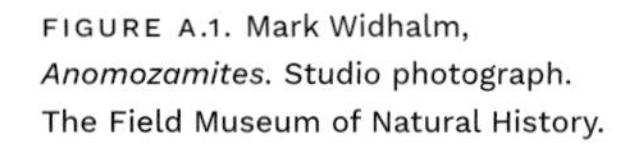
FIGURE A.1. Mark Widhalm, *Anomozamites*. Studio photograph. The Field Museum of Natural History.

overall are a narrow oval in shape (termed *lanceolate*), with the leaf blade divided into feathery segments that are usually as wide as, or wider than they are long. Leaves are also referred to as *paripinnate* because they always end in two (*pari*) segments (*pinna*). The leaf lamina of each segment is attached broadly to the axis (central rib) and has simple or forked veins that run in parallel, predominantly perpendicular to the axis. *Cycadolepis*, the protective wrinkled bracts of proto-flowers, probably belonged to the same plant that bore foliage of the *Anomozamites* type.

Species of *Anomozamites* are identified by the outline and shape of the leaf segments, the density of the veins that supply water to the leaf, and microscopic features that include the shape and arrangement of the stomata, the structure of projections from cells called papillae and leaf hairs, and the outline of the cells of the leaf skin (epidermis). Surprisingly, all of these exquisitely detailed features can be observed on 200 million-year-old fossils! The genus ranges in age from the Late Triassic to the Early Cretaceous, but in Greenland it is characteristic of only the Late Triassic forests.

CLADOPHLEBIS

Cladophlebis is the name for a genus of fossil fern leaves. The genus was first described formally in a publication by French botanist Adolphe Théodore Brongniart in 1849. Counts of fossil specimens from the Astartekløft beds show that *Cladophlebis* was ecologically rare in the forests of both the Triassic and Jurassic, but was highly abundant in the Disaster Bed (representing about 40 percent of the fossils found there).

Cladophlebis was established as a form genus with a fairly vague botanical description. For a fossil to be classified as *Cladophlebis*, it must be a fern leaf (frond) with leaflets (pinnules) attached to a stem (rachis), have a median vein that runs to the apex of each pinnule, and have curved lower-order veins that arise from the median vein and then divide into two equal parts. Excellent fossil preservation and a good hand lens are therefore needed! While it is unclear which evolutionary group of living ferns *Cladophlebis* is most closely related to, some fronds that fit this description have been assigned to the family Osmundaceae, although fossil members of the tropical fern families Dicksoniaceae and Schizaeaceae are also known to have borne *Cladophlebis*-type foliage. However, once a *Cladophlebis*-type frond with attached spore-bearing structures, called sori, is found, it can be identified to family and is typically reclassified. If, as is most commonly the case in compressed fossils, the sori are like those of the Osmundaceae, the frond is typically placed in the genus *Todites*. However, if the anatomy of the frond stem

FIGURE A.2. John Weinstein, *Cladophlebis*. Studio photograph. The Field Museum of Natural History.

is all that is preserved, and it is like that of the Osmundaceae, it is placed in another genus altogether, such as *Osmundites*. Paleobotany is complex, and it is often nearly impossible to know how fossils are related to living plants, if at all, without fossilized reproductive structures such as fern sori or conifer cones. *Cladophlebis*-type morphology exists today. It's simply that, by convention, paleobotanists usually refrain from using form genera for Cenozoic plants (those of the geological era following the K-P mass extinction) and tend to place fossil fronds younger than 66 million years old in genera that live today. Fronds of *Cladophlebis* are so common in some sediments of Mesozoic age that it has been termed the "Mesozoic weed."

CTENIS AND *PSEUDOCTENIS*

Fossil leaves assigned to *Ctenis* and *Pseudoctenis* closely resemble the leaves of modern cycads. They have a very long history in paleobotany, as both genera were established by pioneering British paleobotanists: in 1834, Lindley and Hutton established *Ctenis*, and in 1911, Seward established *Pseudoctenis*. Both have a broad stem with compound leaflets attached along it to form fronds that can measure up to 2 meters in length. Commenting on *Pseudoctenis*, Tom Harris observed that if modern cycad taxa such as *Dioon*, *Macrozamia*, *Encephalartos*, and *Lepidozamia* were to be fossilized, they might be included within this genus. However, there is a high degree of convergence in gross form and leaf venation among different living and extinct cycad genera, so details of their epidermal anatomy are essential to reliably distinguish among them.

Fossils placed in *Ctenis* and *Pseudoctenis* are morphologically very similar at the gross level. In establishing the genus *Ctenis*, Lindley and Hutton emphasized the abundant vein anastomoses, where cross-veins in the pinnules form links between the stronger parallel veins, resulting in a reticulate (netlike) venation pattern. In Seward's mind, leaves that lacked abundant cross venation were sufficiently different to merit placement in a new genus: *Pseudoctenis*. However, epidermal anatomy is still important in the diagnosis of these taxa. Today, more than 50 species, ranging in age from the Late Triassic to the Eocene, have been assigned to the genus *Ctenis*. Fossils assigned to *Pseudoctenis* and *Ctenis* are not common in Greenland, but are found on either side of the Triassic-Jurassic boundary. They generated a lot of excitement when unearthed because of their generally huge size compared with the leaves of other fossil species.

CZEKANOWSKIA

Czekanowskia is a fossil genus established for ribbon-like but overall wedge-shaped leaves that branch dichotomously between one and three times and are borne in bundles on short, fat little shoots covered with persistent scales. These leaves are considered to be seasonally deciduous because they are frequently found in extensive fossilized leaf litter mats, as though the autumn forest floor of Greenland was fossilized. The genus was originally described and formally established by Swiss geologist and naturalist Oswald Heer in 1876, for fossil material of Jurassic age from Russia. Later. Tom Harris and his colleague Jose Miller emended Heer's original description. It was originally assumed that these fossils were evolutionarily related to the ginkgos, but following studies of their fossilized fruits, called *Leptostrobus*, Harris and other authors kept them out of the ginkgo family and in their own enigmatic group, called Czekanowskiales, a group that today has no known relatives. Most researchers still agree with this assessment. In Greenland, *Czekanowskia* fossils are found in extraordinary abundance in the *Spectabilis* and Ledge Beds, suggesting that they were a dominant component of Jurassic forests. Their fossilized feathery leaves were devilishly difficult to bring to life, but Marlene's engineering studies helped us to find a solution.

FIGURE A.3. John Weinstein, *Czekanowskia*. Studio photograph. The Field Museum of Natural History.

FIGURE A.4. John Weinstein, *Dictyophyllum.* Studio photograph. The Field Museum of Natural History.

DICTYOPHYLLUM

Dictyophyllum is an extinct fern genus within the living tropical family Dipteridaceae. It was first described by botanist John Lindley and geologist William Hutton in 1834. Hutton was a great collector of fossil plants found in the coal measures around his home of Sunderland in the north of England. The family Dipteridaceae today is made up of over 20 living species with dissected leaves that have a feather-like appearance called pinnate or pinnatifid. The most characteristic trait of this genus is its leaf venation pattern, which is mesh-like, made up of irregular polygons, rarely rectangles or squares. This genus is very similar in overall shape and structure to *Clathropteris* and *Thaumatopteris*, but is distinguished from them by its general architecture, differences in the sori, and its leaf venation pattern. In a way, the vein patterning of a fern leaf is like the pattern of lines on our own palms and fingers. In the case of humans, every individual person has a distinct pattern that gives us our unique fingerprint. In fern leaves, the distinct patterning of the veins gives each species its unique "fingerprint." *Dictyophyllum* is found throughout the Mesozoic from the Triassic to the Early Cretaceous, and in Greenland it is found on either side of the Triassic-Jurassic boundary, but in higher abundance in Triassic beds.

ELATOCLADUS

Elatocladus is a genus established for a type of fossil leaf that was produced by a conifer often associated with swampy or boggy habitats. It is a minor component of the Greenland vegetation but persisted across the Triassic-Jurassic mass extinction boundary. The genus *Elatocladus* was first described and established in 1913 by renowned Swedish paleobotanist Thore Gustaf Halle for fossil plant material collected from the northwesternmost tip of Antarctica at Hope Bay sometime between 1901 and 1903. The genus was further studied and its formal description updated by English paleobotanist Tom Harris. It is interesting to consider that Harris's study of fossil *Elatocladus* from the Arctic was made possible by initial discoveries and descriptions of *Elatocladus* fossils found in the Antarctic! *Elatocladus* is now typically considered to encompass shoots of conifer trees that bear elongate, flattened needle leaves with a single vein supplying each leaf. The number of leaf veins and their patterning are important traits for classifying plants. In ferns, there are generally multiple veins and a myriad of patterns. In many (but not all) conifers, each leaf contains a single visible vein, which never branches, loops, or forks. The leaves of *Elatocladus* are usually borne in a spirally arranged pattern. One of the most useful characteristics by which to recognize this fossil can be observed at the base of the leaves: at the point where the leaves join the stem, they become constricted to form short little stalks, called petioles. This combination of characters, along with others, is used to distinguish *Elatocladus* from other conifers with similarly shaped leaves, such as the living *Cunninghamia*, which lacks the basal constriction of the needles to form a petiole.

EQUISETITES

Equisetites was first described as a fossil genus by Czech polymath Baron Kaspar Maria von Sternberg—who is considered the father of paleobotany—in 1833. Sternberg used this newly established genus to accommodate all fossil species that had formerly been assigned to the living genus *Equisetum*, which

FIGURE A.5. Mihai Popa, *Equisetites*. Field photograph. Mihai's finger for scale; John Weinstein, *Equisetites*. Studio photograph. The Field Museum of Natural History.

includes modern horsetails and scouring rushes. This change marked a new way of thinking about fossils and helped avoid the pitfalls of shoehorning fossil specimens that looked something (but not exactly) like their presumed living relatives into living genera and species. *Equisetum* is the only living genus in the family Equisetaceae, of the class Polypodiopsida, which also includes the ferns. The horsetails, sometimes known as the sphenopsids or sphenophytes, have one of the longest fossil records of any living group of vascular plants (land plants with specialized tissues for transporting water and minerals around the plant body; this group excludes mosses, hornworts, and liverworts, which lack these tissues). They date back to the Devonian, over 360 million years ago. *Equisetites* reproduce by means of spores instead of seeds. Ferns also reproduce by spores, but the spores of *Equisetites* are especially remarkable. Each spore has four retractable arms that spring open with great force when the spores dry up, which propels the spores like jumping frogs into the air and away from their parent plant. When the spores are wet, the four arms coil back tightly around the spore again in a protective hug. *Equisetites* also have a highly characteristic architecture, in which their stems are divided into a series of nodes (joints) and internodes. Leaves and branches arise from the main stem only at the nodes in a perfect whorl, like ordered spokes on a wheel. Bamboo also has this type of architecture today, with nodes bearing whorls of leaves and/or branches.

FIGURE A.6. Mark Widhalm, *Ginkgoites minuta*. Studio photograph. The Field Museum of Natural History.

GINKGO

Ginkgo was first described by the prolific Swedish botanist Carl Linnaeus in 1771. The genus contains only one living species, *Ginkgo biloba*, which is commonly known as the maidenhair tree. Ginkgos are a highly unusual group of seed plants. They do not have flowers like angiosperms, or cones like conifers and cycads, or spores like ferns. They keep their male and female reproductive structures on separate trees, and their seeds are surrounded by a fleshy and rather stinky "fruit." Fossils assigned to the genus *Ginkgo*/*Ginkgoites* date back to at least the Late Triassic, while the Ginkgoales may have a fossil record extending as far back as the Permian. In Greenland, *Ginkgoites* is found on both sides of the Triassic-Jurassic boundary, but it and other members of the family are much more abundant in post-extinction strata.

In the modern world, *Ginkgo biloba* is endemic to China and is used as a street tree worldwide because of its resistance to urban pollution and temperature extremes. Male trees are preferred by city planners and landscape architects because of the pungent smell of the females' seeds. In the geological past, *Ginkgo* and its relatives were cosmopolitan (distributed across most of the planet), and their fossils have been found on every continent on Earth. The fossils from Greenland, including the one illustrated in this gallery, have leaves that are typically more highly dissected than those of *Ginkgo biloba*, and were assigned by Tom Harris to the species *Ginkgoites minuta*.

LEPIDOPTERIS

Lepidopteris is a genus established for non-fertile foliage that typically comprises a feathery leaf described as *bipinnate*—that is, having two rows of leaflets on opposite sides of a central axis—in which each leaflet is itself composed of leaflets (pinnae) on opposite sides of a central axis (like those of the modern northern wood fern *Dryopteris expansa*). The leaflets are not attached to the central axis opposite each other, but rather are offset and may alternate from one side to the other. The smallest leaflets, called pinnules, are attached across their bases broadly at an angle to the axis, with the edge of each pinnule running into that of the next on either side. The leaves are characterized by the presence of miniature pinnules attached to the main stem, which occur between the larger pinnae and are called intercalary pinnules, having developed after the larger ones. Fossils from Greenland that conform to this very technical description and also have thick cuticles are assignable to *Lepidopteris ottonis*, a woody vine species that went extinct globally at the Triassic-Jurassic boundary.

FIGURE A.7. Mark Widhalm, *Lepidopteris ottonis* fossil peeled off rock. Studio photograph. The Field Museum of Natural History.

Ian recalled in his field notebook from Astartekløft how exhilarating it was to crack the rocks to expose their fossils, particularly those of *Lepidopteris*. Most plants would appear as dark compressions or impressions tightly bound to the rock, but the robust cuticles of *Lepidopteris ottonis* were so superbly preserved that whole fronds were found sitting paperlike on the bedding planes as the rocks were carefully cracked open. But exhilaration could all too soon turn to despair, as those superb delicate fronds were often whipped away by icy, gusting coastal winds at the moment of their revelation—never to be seen again.

FIGURE A.8. John Weinstein, *Nilssonia.* Studio photograph. The Field Museum of Natural History.

NILSSONIA

Nilssonia is a genus for cycad leaf fossils that was first described by French botanist Brongniart in 1825, when he was just 24 years old. The plant to which *Nilssonia* belonged has been interpreted as a woody cycad that was frost resistant and shed its leaves annually in a manner similar to *Ginkgo*. These leaves are generally strap-shaped to oblanceolate (shaped like a lance head, but with the point at the base), with the leaf blade often segmented into variably sized and shaped leaf segments. The venation of the leaf is simple: lower-level veins depart the central vein at right angles. The leaf is hairy on the lower surface, where the stomata are exclusively found. *Nilssonia* ranges in age from Late Triassic to Cretaceous. In Greenland, we interpreted *Nilssonia* as adapted to seasonal flooding because it was found in the Halfway Bed perfectly preserved by a flooding event, which suggested that the leaves had not moved far from the plant they had fallen from. The genus occurs only rarely in East Greenland, but in more Jurassic than Triassic beds.

PODOZAMITES

FIGURE A.9. *Facing:* Mark Widhalm, *Podozamites.* Studio photograph. The Field Museum of Natural History.

FIGURE A.10. *Facing:* John Weinstein, *Pterophyllum.* Studio photograph. The Field Museum of Natural History.

Podozamites is an unusual conifer genus that differs from most living conifers, which have rather narrow leaves and are evergreen. The leaves of *Podozamites* are comparatively large, broad, flattened, and multi-veined; shed on shoots; and were perhaps shed in a deciduous fashion. This last conclusion remains to be tested, however, with further data. *Podozamites* was first described formally as a genus by German paleobotanist Karl Friedrich Wilhelm

Braun in 1843. The genus is characterized as conifer foliage with leaves arranged alternately along the stem (rather than opposite each other), pinched leaf bases, and multiple veins that run parallel to one another for most of the leaf length, like a multilane highway with no junctions or connections.

The genus was widespread across the entire Earth during the Mesozoic. In some parts of the world during the Late Triassic and Early Jurassic, forests were dominated by this genus to the extent that it may have formed almost single-species stands, like those observed in modern commercial forest plantations. There are suggestions that *Podozamites* prospered under warm, humid conditions and that its habit of shoot shedding was an adaptation to heavy predation by insects. The insect predation idea is not supported by the Greenland *Podozamites* fossils, which only very rarely show evidence of insect herbivory and feeding damage. In Greenland, these plants were an important, sometimes dominant, element of the forest canopy up until the Triassic-Jurassic boundary. *Podozamites* survived as a genus, but in very low abundances, in the Jurassic forests once they became reestablished.

PTEROPHYLLUM

Pterophyllum exists as a genus in two kingdoms of life! In the animal kingdom, *Pterophyllum* is a genus of living freshwater fish from the family Cichlidae, perhaps better known as angelfish. However, in the plant kingdom, *Pterophyllum* Brongniart is a genus of leaf fossils similar in appearance to the leaves of some modern cycads, but more particularly the extinct genus *Anomozamites*.

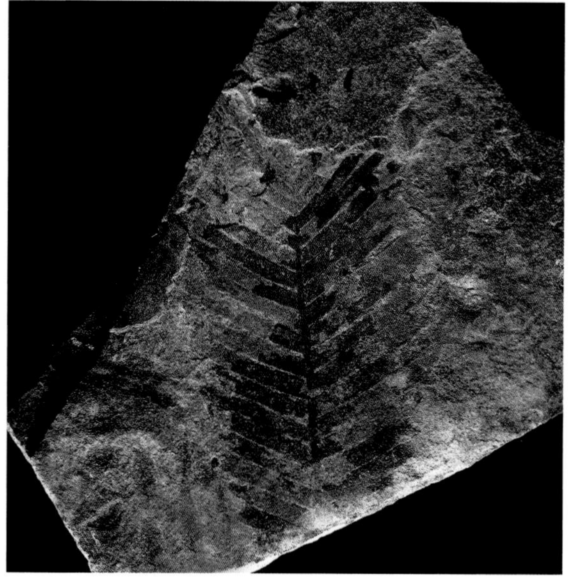

Leaves assigned to *Pterophyllum* are usually regularly segmented; the pinnae are slender and parallel sided. The leaf lamina of each pinna is inserted laterally into the axis, and the parallel veins are simple or forked, running predominantly perpendicular to the axis. Unlike those of *Anomozamites*, the leaves of *Pterophyllum* end with a single pinna, rather than two pinnae (and are therefore referred to as *imparipinnate*). *Pterophyllum* pinnae usually have a width-to-length ratio of 2:1 or less, and the bases of the pinnae are expanded.

Species of *Pterophyllum* are identified by the outline and shape of the pinnae, the number of veins entering each pinna, and microscopic features of the leaves, including the shape, size, and orientation of the stomata. These plants first appeared in the Permian, or perhaps even in the Carboniferous, though these assignments are questionable, but they are most abundant in strata of Late Triassic to Jurassic age. They were an important part of the mid-canopy of the East Greenland forests prior to the Triassic-Jurassic mass extinction, and although they survived it, they were present in very low abundances in the Jurassic forests.

SPHENOBAIERA

Sphenobaiera is a fossil genus first described and established by German botanist Rudolph Florin in 1936. *Sphenobaiera* includes all leaf fossils that are deeply dissected and have an overall wedge shape. If you look carefully at the leaves of *Sphenobaiera*, you will see multiple veins that run along the leaf and then fork into two veins (dichotomous venation) in a manner reminiscent of living *Ginkgo biloba*. *Sphenobaiera* can be distinguished from leaf fossils of *Ginkgo* and *Baiera* by the absence of a petiole. While fossils assigned to *Sphenobaiera* are widely considered to belong to the ginkgo family, this relationship has never been conclusively demonstrated, and these fossils remain enigmatic. In Greenland, these fossils became a significant element of the vegetation only following the Triassic-Jurassic boundary; they are particularly prolific in the *Spectabilis* and Ledge Beds of Astartekløft, where we believe they were large trees that dominated the canopy.

STACHYOTAXUS

Stachyotaxus was first described by Swedish geologist and botanist Alfred Gabriel Nathorst in 1886. The genus was established to include conifer leafy shoots and cones in which the proximal part of the shoot bears small, scale-like leaves while distally, the leaves are shaped like a sword blade (ensiform). The cones are loose, spirally arranged, bract-scale complexes, each bearing a single seed in a cuplike structure on the upper surface. *Stachyotaxus* was highly abundant in the Boundary Bed at Astartekløft. This genus is known only from the latest stage of the Triassic—the Rhaetian—and from no other time in Earth history, which is intriguing and warrants further study.

FIGURE A.11. Mark Widhalm, *Sphenobaiera spectabilis.* Studio photograph. The Field Museum of Natural History.

FIGURE A.12. John Weinstein, *Stachyotaxus* cone. Studio photograph. The Field Museum of Natural History.

Further Reading

1. A JOURNEY INTO THE PAST

Behrensmeyer, Anna K., Susan M. Kidwell, and Robert A. Gastaldo. "Taphonomy and paleobiology." *Paleobiology* 26, no. S4 (2000): 103–147.

Campbell, David G., Judy L. Stone, and A. R. I. T. O. Rosas Jr. "A comparison of the phytosociology and dynamics of three floodplain (várzea) forests of known ages, Rio Juruá, western Brazilian Amazon." *Botanical Journal of the Linnean Society* 108, no. 3 (1992): 213–237.

Cohen, K. M., S. C. Finney, P. L. Gibbard, and J.-X. Fan. "The ICS international chronostratigraphic chart (2013 updated)." *Episodes* 36, no. 3 (2013): 199–204.

Dam, Gregers, and Finn Surlyk. "Forced regressions in a large wave- and storm-dominated anoxic lake, Rhaetian-Sinemurian Kap Stewart Formation, East Greenland." *Geology* 20, no. 8 (1992): 749–752.

Dam, Gregers, and Finn Surlyk. "Sequence stratigraphic correlation of Lower Jurassic shallow marine and paralic successions across the Greenland-Norway Seaway." In *Sequence stratigraphy on the northwest European margin*, edited by R. J. Steel, V. L. Felt, E. P. Johannesson, and C. Mathieu, 483–509. NPF Special Publication 5. Elsevier, 1995.

Dam, Gregers, Finn Surlyk, H. W. Posamentier, C. P. Summerhayes, B. U. Haq, and G. P. Allen. "Cyclic sedimentation in a large wave- and storm-dominated anoxic lake; Kap Stewart Formation (Rhaetian-Sinemurian), Jameson Land, East Greenland." In *Sequence stratigraphy and facies associations*, edited by G. Dam, F. Surlyk, H. W. Posamentier, C. P. Summerhayes, B. U. Haq, and G. P. Allen, 419–448. IAS Special Publication 18. Blackwell Scientific, 1993.

Harris, Thomas Maxwell. "The fossil flora of Scoresby Sound East Greenland." Parts 1–5. *Meddelelser om Gronland* (1926–1937).

Hartz, Nikolaj. "Planteforsteninger fra Cap Stewart i Østgrønland, med en historisk oversigt." *Bianco Lunos Kgl. Hof-Bogtrykkeri*, 1896.

Holland, Marika M., and Cecilia M. Bitz. "Polar amplification of climate change in coupled models." *Climate Dynamics* 21, nos. 3–4 (2003): 221–232.

McElwain, Jennifer C., David J. Beerling, and F. Ian Woodward. "Fossil plants and global warming at the Triassic-Jurassic boundary." *Science* 285, no. 5432 (1999): 1386–1390.

McElwain, Jennifer C., Mihai E. Popa, Stephen P. Hesselbo, Matthew Haworth, and Finn Surlyk. "Macroecological responses of terrestrial vegetation to climatic and atmospheric change across the Triassic/Jurassic boundary in East Greenland." *Paleobiology* 33, no. 4 (2007): 547–573.

McElwain, Jennifer C., and Surangi W. Punyasena. "Mass extinction events and the plant fossil record." *Trends in Ecology & Evolution* 22, no. 10 (2007): 548–557.

McElwain, Jennifer C., Peter J. Wagner, and Stephen P. Hesselbo. "Fossil plant relative abundances indicate sudden loss of Late Triassic biodiversity in East Greenland." *Science* 324, no. 5934 (2009): 1554–1556.

Nichols, Gary. *Sedimentology and stratigraphy*. John Wiley & Sons, 2009.

Stamp, Tom, and Cordelia Stamp. *William Scoresby, Arctic scientist*. Caedmon of Whitby Press, 1976.

2. FORESTS OF A LOST LANDSCAPE

Arnold, Sarah E., Samia Faruq, Vincent Savolainen, Peter W. McOwan, and Lars Chittka. "FReD: the floral reflectance database—a web portal for analyses of flower colour." *PLoS One* 5, no. 12 (2010): e14287.

Behrensmeyer, Anna K., Susan M. Kidwell, and Robert A. Gastaldo. "Taphonomy and paleobiology." *Paleobiology* 26, no. S4 (2000): 103–147.

Crane, Peter R. *Ginkgo: the tree that time forgot*. Yale University Press, 2013.

Enquist, Brian J., and Karl J. Niklas. "Invariant scaling relations across tree-dominated communities." *Nature* 410, no. 6829 (2001): 655.

Friis, Else Marie, Peter R. Crane, and Kaj Raunsgaard Pedersen. *Early flowers and angiosperm evolution*. Cambridge University Press, 2011.

Gatesy, Stephen M., Kevin M. Middleton, Farish A. Jenkins Jr., and Neil H. Shubin. "Three-dimensional preservation of foot movements in Triassic theropod dinosaurs." *Nature* 399, no. 6732 (1999): 141–144.

Harris, Thomas Maxwell. *The Yorkshire Jurassic flora*. III. *Bennettitales*. British Museum (Natural History), 1969.

Jones, Hamlyn G. *Plants and microclimate: a quantitative approach to environmental plant physiology*. Cambridge University Press, 2013.

Labandeira, Conrad C. "A paleobiologic perspective on plant-insect interactions." *Current Opinion in Plant Biology* 16, no. 4 (2013): 414–421.

McElwain, Jennifer C., and Margret Steinthorsdottir. "Paleoecology, ploidy, paleoatmospheric composition, and developmental biology: a review of the multiple uses of fossil stomata." *Plant Physiology* 174, no. 2 (2017): 650–664.

Nadkarni, Nalini M., Geoffrey G. Parker, H. Bruce Rinker, and David M. Jarzen. "The nature of forest canopies." In *Forest Canopies*, 2nd ed., edited by Margaret D. Lowman and H. Bruce Rinker, 3–23. Academic Press, 2004.

Page, Christopher N., and Patrick J. Brownsey. "Tree-fern skirts: a defence against climbers and large epiphytes." *Journal of Ecology* (1986): 787–796.

Schnitzer, Stefan, Frans Bongers, Robyn J. Burnham, and Francis E. Putz, eds. *Ecology of lianas*. John Wiley & Sons, 2014.

Soh, Wuu Kuang, Ian J. Wright, Karen L. Bacon, Tanja I. Lenz, Margret Steinthorsdottir, Andrew C. Parnell, and Jennifer C. McElwain. "Palaeo leaf economics reveal a shift in ecosystem function associated with the end-Triassic mass extinction event." *Nature Plants* 3, no. 8 (2017): 17104.

Steinthorsdottir, Margret, Andrew J. Jeram, and Jennifer C. McElwain. "Extremely elevated CO_2 concentrations at the Triassic/Jurassic boundary." *Palaeogeography, Palaeoclimatology, Palaeoecology* 308, nos. 3–4 (2011): 418–432.

Steinthorsdottir, Margret, Anne-Marie P. Tosolini, and Jennifer C. McElwain. "Evidence for insect and annelid activity across the Triassic-Jurassic transition of East Greenland." *Palaios* 30, no. 8 (2015): 597–607.

Steinthorsdottir, Margret, F. Ian Woodward, Finn Surlyk, and Jennifer C. McElwain. "Deep-time evidence of a link between elevated CO_2 concentrations and perturbations in the hydrological cycle via drop in plant transpiration." *Geology* 40, no. 9 (2012): 815–818.

Taylor, Edith L., Thomas N. Taylor, and Michael Krings. *Paleobotany: the biology and evolution of fossil plants*. Academic Press, 2009.

Whalley, Paul. "Insects from the Italian Upper Trias." *Rivista del Museo Civico di Scienze Naturali "Enrico Caffi" Bergamo* 10 (1986): 51–60.

Wright, Ian J., Peter B. Reich, Mark Westoby, David D. Ackerly, Zdravko Baruch, Frans Bongers, Jeannine Cavender-Bares et al. "The worldwide leaf economics spectrum." *Nature* 428, no. 6985 (2004): 821.

Zhou, Ning, Yong-Dong Wang, Li-Qin Li, and Xiao-Qing Zhang. "Diversity variation and tempo-spatial distributions of the Dipteridaceae ferns in the Mesozoic of China." *Palaeoworld* 25, no. 2 (2016): 263–286.

3. CRISIS AND COLLAPSE

Bacon, Karen L., Claire M. Belcher, Matthew Haworth, and Jennifer C. McElwain. "Increased atmospheric SO_2 detected from changes in leaf physiognomy across the Triassic-Jurassic boundary interval of East Greenland." *PLoS One* 8, no. 4 (2013): e60614.

Bacon, Karen L., Claire M. Belcher, Stephen P. Hesselbo, and Jennifer C. McElwain. "The Triassic-Jurassic boundary carbon-isotope excursions expressed in taxonomically identified leaf cuticles." *Palaios* 26, no. 8 (2011): 461–469.

Belcher, Claire M., Luke Mander, Guillermo Rein, Freddy X. Jervis, Matthew Haworth, Stephen P. Hesselbo, Ian J. Glasspool, and Jennifer C. McElwain. "Increased fire activity at the Triassic/Jurassic boundary in Greenland due to climate-driven floral change." *Nature Geoscience* 3, no. 6 (2010): 426.

Bowman, David M. J. S., Jennifer K. Balch, Paulo Artaxo, William J. Bond, Jean M. Carlson, Mark A. Cochrane, Carla M. D'Antonio et al. "Fire in the Earth system." *Science* 324, no. 5926 (2009): 481–484.

Cochrane, Mark A., Ane Alencar, Mark D. Schulze, Carlos M. Souza, Daniel C. Nepstad, Paul Lefebvre, and Eric A. Davidson. "Positive feedbacks in the fire dynamic of closed canopy tropical forests." *Science* 284, no. 5421 (1999): 1832–1835.

Cope, M. J., and W. G. Chaloner. "Fossil charcoal as evidence of past atmospheric composition." *Nature* 283, no. 5748 (1980): 647.

Elliott-Kingston, Caroline, Matthew Haworth, and Jennifer C. McElwain. "Damage structures in leaf epidermis and cuticle as an indicator of elevated atmospheric sulphur dioxide in early Mesozoic floras." *Review of Palaeobotany and Palynology* 208 (2014): 25–42.

Flessa, Karl W., and David Jablonski. "Extinction is here to stay." *Paleobiology* 9, no. 4 (1983): 315–321.

Glasspool, Ian J., and Andrew C. Scott. "Identifying past fire events." *Fire phenomena and the earth system: an interdisciplinary guide to fire science* (2013): 177–206.

Grattan, John, Roland Rabartin, Stephen Self, and Thorvaldur Thordarson. "Volcanic air pollution and mortality in France, 1783–1784." *Comptes Rendus Geoscience* 337, no. 7 (2005): 641–651.

Haworth, Matthew, Caroline Elliott-Kingston, Angela Gallagher, Annmarie Fitzgerald, and Jennifer C. McElwain. "Sulphur dioxide fumigation effects on stomatal density and index of non-resistant plants: implications for the stomatal palaeo-[CO_2] proxy method." *Review of Palaeobotany and Palynology* 182 (2012): 44–54.

Haworth, Matthew, Angela Gallagher, Elysia Sum, Marlene Hill Donnelly, Margret Steinthorsdottir, and Jennifer McElwain. "On the reconstruction of plant photosynthetic and stress physiology across the Triassic-Jurassic boundary." *Turkish Journal of Earth Sciences* 23, no. 3 (2014): 321–329.

Hesselbo, Stephen P., Jennifer C. McElwain, Mihai Popa, Finn Surlyk, and Matthew Haworth. "New floral, sedimentological and isotopic investigation of the Triassic-Jurassic boundary strata in Jameson Land, East Greenland." In *3rd Workshop on the IGCP Project*, vol. 458 (2003): 22–23.

Hesselbo, Stephen P., Stuart A. Robinson, Finn Surlyk, and Stefan Piasecki. "Terrestrial and marine extinction at the Triassic-Jurassic boundary synchronized with major carbon-cycle perturbation: a link to initiation of massive volcanism?" *Geology* 30, no. 3 (2002): 251–254.

Lindström, Sofie, Hamed Sanei, Bas Van De Schootbrugge, Gunver K. Pedersen, Charles E. Lesher, Christian Tegner, Carmen Heunisch et al. "Volcanic mercury and mutagenesis in land plants during the end-Triassic mass extinction." *Science Advances* 5, no. 10 (2019): eaaw4018.

Mander, Luke. "Taxonomic resolution of the Triassic-Jurassic sporomorph record in East Greenland." *Journal of Micropalaeontology* 30, no. 2 (2011): 107–118.

Mander, Luke, Wolfram M. Kürschner, and Jennifer C. McElwain. "An explanation for conflicting records of Triassic-Jurassic plant diversity." *Proceedings of the National Academy of Sciences* 107, no. 35 (2010): 15351–15356.

Mander, Luke, Wolfram M. Kürschner, and Jennifer C. McElwain. "Palynostratigraphy and vegetation history of the Triassic-Jurassic transition in East Greenland." *Journal of the Geological Society* 170, no. 1 (2013): 37–46.

Mander, Luke, Cassandra J. Wesseln, Jennifer C. McElwain, and Surangi W. Punyasena. "Tracking taphonomic regimes using chemical and mechanical damage of pollen and spores: an example from the Triassic-Jurassic mass extinction." *PLoS One* 7, no. 11 (2012): e49153.

Marzoli, Andrea, Paul R. Renne, Enzo M. Piccirillo, Marcia Ernesto, Giuliano Bellieni, and Angelo De Min. "Extensive 200 million-year-old continental flood basalts of the Central Atlantic Magmatic Province." *Science* 284, no. 5414 (1999): 616–618.

McElwain, Jennifer C., Mihai E. Popa, Stephen P. Hesselbo, Matthew Haworth, and Finn Surlyk. "Macroecological responses of terrestrial vegetation to climatic and atmospheric change across the Triassic/Jurassic boundary in East Greenland." *Paleobiology* 33, no. 4 (2007): 547–573.

McElwain, Jennifer C., Peter J. Wagner, and Stephen P. Hesselbo. "Fossil plant relative abundances indicate sudden loss of Late Triassic biodiversity in East Greenland." *Science* 324, no. 5934 (2009): 1554–1556.

Pedersen, K. Raunsgaard, and Jens J. Lund. "Palynology of the plant-bearing Rhaetian to Hettangian Kap Stewart Formation, Scoresby Sund, East Greenland." *Review of Palaeobotany and Palynology* 31 (1980): 1–69.

Pyne, Stephen J. *Introduction to wildland fire. Fire management in the United States*. John Wiley & Sons, 1984.
Robinson, Jennifer M. "Phanerozoic atmospheric reconstructions: a terrestrial perspective." *Palaeogeography, Palaeoclimatology, Palaeoecology* 97, nos. 1–2 (1991): 51–62.
Scott, Andrew C. "The pre-Quaternary history of fire." *Palaeogeography, Palaeoclimatology, Palaeoecology* 164, nos. 1–4 (2000): 281–329.
Scott, Andrew C., and Ian J. Glasspool. "The diversification of Paleozoic fire systems and fluctuations in atmospheric oxygen concentration." *Proceedings of the National Academy of Sciences* 103, no. 29 (2006): 10861–10865.
Surlyk, Finn, Jennifer C. McElwain, Stephen P. Hesselbo, Stefan Piasecki, Mihai E. Popa, and Stuart Robinson. "Macrofloral turnover and anatomical-morphological changes, atmospheric CO_2 increase, super greenhouse conditions and isotope excursions across the Triassic-Jurassic boundary in East Greenland." In *International symposium: Mesozoic-Cenozoic bioevents*, Berlin (2002): 47–49.
Van de Schootbrugge, Bas, Tracy T. M. Quan, Sofie Lindström, Wilhelm Püttmann, Carmen Heunisch, Jörg Pross, Jens Fiebig et al. "Floral changes across the Triassic/Jurassic boundary linked to flood basalt volcanism." *Nature Geoscience* 2, no. 8 (2009): 589–594.
Whelan, Robert J. *The ecology of fire*. Cambridge University Press, 1995.
Williford, Kenneth H., Kliti Grice, Alexander Holman, and Jennifer C. McElwain. "An organic record of terrestrial ecosystem collapse and recovery at the Triassic-Jurassic boundary in East Greenland." *Geochimica et Cosmochimica Acta* 127 (2014): 251–263.
Witton, Mark P. *Pterosaurs: natural history, evolution, anatomy*. Princeton University Press, 2013.

4. RECOVERY OF A TROPICAL ARCTIC

Bacon, Karen L., Claire M. Belcher, Stephen P. Hesselbo, and Jennifer C. McElwain. "The Triassic-Jurassic boundary carbon-isotope excursions expressed in taxonomically identified leaf cuticles." *Palaios* 26, no. 8 (2011): 461–469.
Barnosky, Anthony D., Nicholas Matzke, Susumu Tomiya, Guinevere O. U. Wogan, Brian Swartz, Tiago B. Quental, Charles Marshall et al. "Has the Earth's sixth mass extinction already arrived?" *Nature* 471, no. 7336 (2011): 51–57.
Crane, Peter R., *Ginkgo: the tree that time forgot*. Yale University Press, 2013.
Dam, Gregers, and Finn Surlyk. "Forced regressions in a large wave- and storm-dominated anoxic lake, Rhaetian-Sinemurian Kap Stewart Formation, East Greenland." *Geology* 20, no. 8 (1992): 749–752.
Dam, Gregers, and Finn Surlyk. "Sequence stratigraphic correlation of Lower Jurassic shallow marine and paralic successions across the Greenland-Norway Seaway." In *Sequence stratigraphy on the northwest European margin*, edited by R. J. Steel, V. L. Felt, E. P. Johannesson, and C. Mathieu, 483–509. NPF Special Publication 5. Elsevier, 1995.
Dam, Gregers, Finn Surlyk, H. W. Posamentier, C. P. Summerhayes, B. U. Haq, and G. P. Allen. "Cyclic sedimentation in a large wave- and storm-dominated anoxic lake; Kap Stewart Formation (Rhaetian-Sinemurian), Jameson Land, East Greenland." In *Sequence stratigraphy and facies associations*, edited by G. Dam, F. Surlyk, H. W. Posamentier, C. P. Summerhayes, B. U. Haq, and G. P. Allen, 419–448. IAS Special Publication 18. Blackwell Scientific, 1993.
Flessa, Karl W., and David Jablonski. "Extinction is here to stay." *Paleobiology* 9, no. 4 (1983): 315–321.

Hallam, Anthony, and Paul B. Wignall. *Mass extinctions and their aftermath*. Oxford University Press, 1997.

Harris, Thomas M., and Jose Miller. *The Yorkshire Jurassic Flora. IV. Czekanowskiales*. British Museum (Natural History), 1974.

Krause, Bernie. *Voices of the wild: animal songs, human din, and the call to save natural soundscapes*. Yale University Press, 2015.

Kürschner, Wolfram M. "The anatomical diversity of recent and fossil leaves of the durmast oak (*Quercus petraea* Lieblein/*Q. pseudocastanea* Goeppert): implications for their use as biosensors of palaeoatmospheric CO_2 levels." *Review of Palaeobotany and Palynology* 96 (1997): 1–30.

Kürschner, Wolfram M., Luke Mander, and Jennifer C. McElwain. "A gymnosperm affinity for *Ricciisporites tuberculatus* Lundblad: implications for vegetation and environmental reconstructions in the Late Triassic." *Palaeobiodiversity and Palaeoenvironments* 94, no. 2 (2014): 295–305.

McElwain, Jennifer C., and Surangi W. Punyasena. "Mass extinction events and the plant fossil record." *Trends in Ecology & Evolution* 22, no. 10 (2007): 548–557.

McElwain, Jennifer C., and Margret Steinthorsdottir. "Paleoecology, ploidy, paleoatmospheric composition, and developmental biology: a review of the multiple uses of fossil stomata." *Plant Physiology* 174, no. 2 (2017): 650–664.

Miner, Ernest L. "Paleobotanical examinations of Cretaceous and Tertiary coals." *American Midland Naturalist* (1935): 585–625.

Nathorst, Alfred Gabriel. "Om nógra *Ginkgo* vöxter fran Kolgrufvorna vid Stabbarp i Skåne." *Acta Universitatis Lundensis*, Series 2, no. 2 (1906): 1–15.

Nichols, Gary. *Sedimentology and stratigraphy*. John Wiley & Sons, 2009.

Prieto-Torres, David A., Andres M. Cuervo, and Elisa Bonaccorso. "On geographic barriers and Pleistocene glaciations: tracing the diversification of the russet-crowned warbler (*Myiothlypis coronata*) along the Andes." *PLoS One* 13, no. 3 (2018).

Wotzlaw, Jörn-Frederik, Jean Guex, Annachiara Bartolini, Yves Gallet, Leopold Krystyn, Christopher A. McRoberts, David Taylor, Blair Schoene, and Urs Schaltegger. "Towards accurate numerical calibration of the Late Triassic: High-precision U-Pb geochronology constraints on the duration of the Rhaetian." *Geology* 42, no. 7 (2014): 571–574.

APPENDIX: A FOSSIL PLANT GALLERY

Braun, Karl Friedrich Wilhelm. *Beiträge zur Urgeschichte der Pflanzen*. Vol. 1. Gedruckt bei FC Birner, 1843.

Brongniart, Adolphe. *Histoire des végétaux fossiles, ou recherches botaniques et géologiques*. Vol. 1. Masson, 1836.

Brongniart, Adolphe. *Observations sur les végétaux fossiles renfermés dans les grès de Hoer en Scanie*. 1825.

Brongniart, Adolphe. *Tableau des genres de végétaux fossiles considérés sous le point de vue de leur classification botanique et de leur distribution géologique*. Imprimerie de L. Martinet, 1849.

Cantrill, David J. "Broad leafed coniferous foliage from the Lower Cretaceous Otway Group, southeastern Australia." *Alcheringa* 15, no. 3 (1991): 177–190.

Chen, Li-Qun, Cheng-Sen Li, William G. Chaloner, David J. Beerling, Qi-Gao Sun, Margaret E. Collinson, and Peter L. Mitchell. "Assessing the potential for the stomatal characters of extant and fossil *Ginkgo* leaves to signal atmospheric CO_2 change." *American Journal of Botany* 88, no. 7 (2001): 1309–1315.

Erdei, Boglárka, and Steven R. Manchester. "*Ctenis clarnoensis* sp. n., an unusual cycadalean foliage from the Eocene Clarno Formation, Oregon." *International Journal of Plant Sciences* 176, no. 1 (2014): 31–43.

Florin, Rudolf. "Die fossilen Ginkgophyten von Franz-Joseph-Land nebst Erörterungen über vermeintliche Cordaitales mesozoischen Alters. I. Spezieller Teil." *Palaeontographica Abteilung B* (1936): 71–173.

Halle, Thore Gustaf. *The Mesozoic flora of Graham Land*. Lithographisches Institut des Generalstabs, 1913.

Harris, Thomas Maxwell. "The fossil flora of Scoresby Sound East Greenland." Parts 1–5. *Meddelelser om Gronland* (1926–1937).

Harris, Thomas Maxwell. *The Yorkshire Jurassic Flora*. II. *Caytoniales, Cycadales and Pteridosperms*. British Museum (Natural History), 1964.

Harris, Thomas Maxwell. *The Yorkshire Jurassic Flora*. III. *Bennettitales*. British Museum (Natural History), 1969.

Harris, Thomas Maxwell. *The Yorkshire Jurassic Flora*. V. *Coniferales*. British Museum (Natural History), 1979.

Harris, Thomas M., and Jose Miller. *The Yorkshire Jurassic Flora*. IV. *Czekanowskiales*. British Museum (Natural History), 1974.

Heer, Oswald. *Beiträge zur Jura-flora Ostsibiriens und des Amurlandes*. Vol. 4. Académie Impériale des sciences, 1876.

Lindley, John, and William Hutton. *The fossil flora of Great Britain*. Ridgway, 1833–1835.

Linnaeus, Carolus. *Mantissa plantarum altera generum editionis VI et specierum editionis II*. impensis direct. Laurentii Salvii, 1771.

McLoughlin, Stephen, Raymond J. Carpenter, and Christian Pott. "*Ptilophyllum muelleri* (Ettingsh.) comb. nov. from the Oligocene of Australia: last of the Bennettitales?" *International Journal of Plant Sciences* 172, no. 4 (2011): 574–585.

Mildenhall, Dallas C. "The record of the rocks." *New Zealand's Nature Heritage* 1 (1974): 43–47.

Nathorst, Alfred Gabriel. *Om floran i Skånes kolförande bildningar*. PA Norstedt & söner, kongl. boktryckare, 1886.

Nathorst, Alfred Gabriel. "Om nógra *Ginkgo* vöxter fran Kolgrufvorna vid Stabbarp i Skåne." *Acta Universitatis Lundensis*, Series 2, no. 2 (1906): 1–15.

Pant, Divya Darshan. "The classification of gymnospermous plants." *Palaeobotanist* 6, no. 2 (1957): 65–70.

Pole, Mike, Yong-Dong Wang, Eugenia V. Bugdaeva, Chong Dong, Ning Tian, Li-Qin Li, and Ning Zhou. "The rise and demise of *Podozamites* in east Asia—an extinct conifer life style." *Palaeogeography, Palaeoclimatology, Palaeoecology* 464 (2016): 97–109.

Popa, Mihai E., Ovidiu Barbu, and Vlad Codrea. "Aspects of Romanian Early Jurassic palaeobotany and palynology. Part V. *Thaumatopteris brauniana* from Şuncuiuş." *Acta Palaeontologica Romaniae* 4 (2003): 361–367.

Popa, Mihai E., and Jennifer C. McElwain. "Bipinnate *Ptilozamites nilssonii* from Jameson Land and new considerations on the genera *Ptilozamites* Nathorst 1878 and *Ctenozamites* Nathorst 1886." *Review of Palaeobotany and Palynology* 153, nos. 3–4 (2009): 386–393.

Potonie, Robert. "Synopsis der Gattungen der Sporae dispersae. Teil 1: Sporites." *Geologischen Jahrbuch* 23 (1956): 1–103.

Pott, Christian, Hans Kerp, and Michael Krings. "Morphology and epidermal anatomy of *Nilssonia* (cycadalean foliage) from the Upper Triassic of Lunz (Lower Austria)." *Review of Palaeobotany and Palynology* 143, nos. 3–4 (2007): 197–217.

Pott, Christian, and Michael Krings. "Gymnosperm foliage from the Upper Triassic of Lunz, Lower Austria: an annotated check list and identification key." *Geo. Alp* 7 (2010): 19–38.

Pott, Christian, Stephen McLoughlin, and Anna Lindström. "Late Palaeozoic foliage from China displays affinities to Cycadales rather than to Bennettitales necessitating a re-evaluation of the Palaeozoic *Pterophyllum* species." *Acta Palaeontologica Polonica* 55, no. 1 (2009): 157–168.

Schimper, Wilhelm Philip. *Traité de paléontologie végétale: ou, La flore du monde primitif dans ses rapports avec les formations géologiques et la flore du monde actuel.* Vols. 1–3. JB Baillière et fils, 1869–1870.

Seward, Albert Charles. "XXIII.—The Jurassic Flora of Sutherland." *Earth and Environmental Science Transactions of The Royal Society of Edinburgh* 47, no. 4 (1911): 643–709.

Steinthorsdottir, Margret, Karen L. Bacon, Mihai E. Popa, Laura Bochner, and Jennifer C. McElwain. "Bennettitalean leaf cuticle fragments (here *Anomozamites* and *Pterophyllum*) can be used interchangeably in stomatal frequency-based palaeo-CO_2 reconstructions." *Palaeontology* 54, no. 4 (2011): 867–882.

Sternberg, Kaspar Maria. "1838. Versuch einer geognostisch-botanischen Darstellung der Flora der Vorwelt." In *Kommission im Deutschen Museum, in Leipzig bei Fr. Fleischer* (1820), Fasc. 2.

Taylor, Edith L., Thomas N. Taylor, and Michael Krings. *Paleobotany: the biology and evolution of fossil plants.* Academic Press, 2009.

Turland, Nick J., John Harry Wiersema, Fred R. Barrie, Werner Greuter, David L. Hawksworth, Patrick Stephen Herendeen, Sandra Knapp et al. *International Code of Nomenclature for algae, fungi, and plants (Shenzhen Code) adopted by the Nineteenth International Botanical Congress Shenzhen, China, July 2017.* Koeltz Botanical Books, 2018.

van Konijnenburg-van Cittert, Johanna H. A., Christian Pott, Christopher J. Cleal, and Gea Zijlstra. "Differentiation of the fossil leaves assigned to *Taeniopteris*, *Nilssoniopteris* and *Nilssonia* with a comparison to similar genera." *Review of Palaeobotany and Palynology* 237 (2017): 100–106.

Webb, John A. "Triassic species of *Dictyophyllum* from eastern Australia." *Alcheringa* 6, no. 2 (1982): 79–91.

Zhou, Ning, Yong-Dong Wang, Li-Qin Li, and Xiao-Qing Zhang. "Diversity variation and tempo-spatial distributions of the Dipteridaceae ferns in the Mesozoic of China." *Palaeoworld* 25, no. 2 (2016): 263–286.

Index

Note: Page numbers followed by *f* include a figure.
For a visual index to the species included in the book's landscape reconstructions, please refer to https://press.uchicago.edu/sites/mcelwain/index.html.

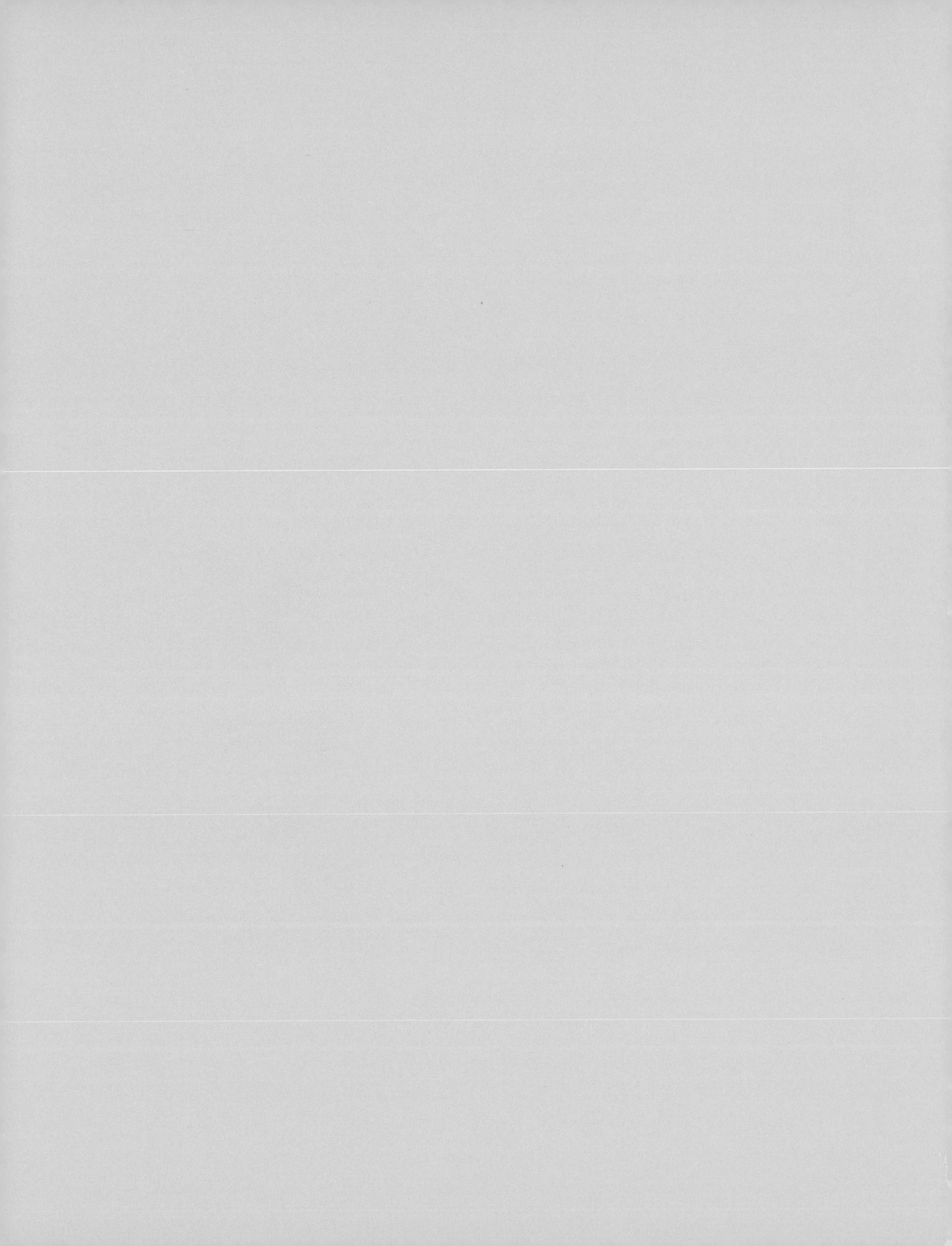